EXAM *Revision* NOTES

AS/A-LEVEL

Physics

Eric Webster

2nd Edition

Philip Allan Updates, an imprint of Hodder Education, part of Hachette Livre UK, Market Place, Deddington, Oxfordshire OX15 0SE

Orders
Bookpoint Ltd, 130 Milton Park, Abingdon, Oxfordshire OX14 4SB
tel: 01235 827720
fax: 01235 400454
e-mail: uk.orders@bookpoint.co.uk

Lines are open 9.00 a.m.–5.00 p.m., Monday to Saturday, with a 24-hour message answering service. You can also order through the Philip Allan Updates website: www.philipallan.co.uk

ISBN 978-0-340-95861-2

First printed 2008
Impression number 5 4 3 2 1
Year 2013 2012 2011 2010 2009 2008

Printed in Spain

P01268

Contents

Introduction

These revision notes cover all the material in the core AS and A-level physics specifications of a number of examination boards. While every attempt has been made to make the notes comprehensive, you should check with the specifications of the board you are taking to see which specification topics you require.

Each topic is broken down into sections which are designed to stand alone and, at the end of most sections, three worked examples are provided followed by a 'You should now know' list. If you feel that you know the material in a given section, a glance at the worked questions and the 'You should now know' list will enable you to decide quickly whether you can skip the topic, or whether it is necessary to work through the material.

This introductory section deals with some general points that you need to consider when undertaking the AS and A-level examination. It also contains details of what is expected when questions are asked in a particular format. I have included some hints, based upon my experience as an examiner, to enable you to obtain the maximum marks possible.

1 *Significant figures*

With the advent of calculators, the problem of significant figures used in calculations has grown in importance. Depending on the calculator you use, it is possible to give answers to eight or perhaps ten figures, even when the data in the question are only given to three figures. Clearly, in such cases, some of the numbers obtained are pure chance, depending on how the calculation is performed. As scientists, we need to give numerical answers that contain only numbers that are significant to the answer.

Most examination calculations give values to be used in a calculation to two or three significant figures. A simple rule to apply is that the answer should be quoted to the same number of significant figures as given in the question. If there is a mixture of numbers in the question, then your answer should be quoted to the same number of significant figures as the quantity with the least number of significant figures.

Some examination boards will allow answers quoted to one significant figure above or one below the correct number without penalty, but use this simple rule to be on the safe side.

Example

Calculate the gravitational force on the Moon due to the Earth. The Earth has a mass of 5.98×10^{24} kg, the mass of the Moon is 7.35×10^{22} kg, the distance from the Moon to the Earth is 3.84×10^{8} m and the gravitational constant $G = 6.672 \times 10^{-11}$ N m^2 kg^{-2}.

Since some values are given to three significant figures and one to four significant figures, you should quote your answer to the smallest number of significant figures given, in this case three.

$$F = \frac{Gm_1m_2}{d^2} = \frac{6.672 \times 10^{-11} \times 5.98 \times 10^{24} \times 7.35 \times 10^{22}}{(3.84 \times 10^{8})^2}$$

$$F = 1.988\ldots \times 10^{20} = 1.99 \times 10^{20}\text{ N}$$

Caution Do not round figures down early in the calculation, as this can generate significant errors, which increase as the calculation proceeds. As shown in the above example, you should work with the figures given, use the four given for G and round down when you get to the answer.

1.1 Rounding down rules

The rule to round down is that if the number you need to remove is in the range 0 to 4, do not add 1 to the previous number; if it is in the range 5 to 9, add 1 to the previous number.

Rounding to three figures

2 Units

When quoting an answer to a calculation, you should include the correct units. There is no need to convert the units to base units; conventional derived units are acceptable.

No units or wrong units will almost certainly lose a mark.

Example

In the gravity question, the units of force can be quoted as N or $kg\,m\,s^{-2}$.

3 Pre-printed answer books

For certain papers, some examination boards provide pre-printed books for your answers. If a calculation is to be performed, there will be a space for the answer. Sometimes the units will be given and you must calculate your answer to match the units given.

Example

You have been asked to calculate the wavelength of green light. You obtain a value 5.53×10^{-7} m. If the place where your answer is to be written already gives the units as nm, the answer required is 553. So be on your guard: you would lose a mark for writing 5.53×10^{-7} as your answer.

4 General points to consider when performing calculations

- Always show your working, and try to keep it as neat as possible. The examiner needs to be able to follow it. Don't just give the answer that you have worked out on a separate sheet.
- Write down equations using the usual symbols. The examiner will be familiar with the equation required and, when you substitute values, should be able to follow your working.
- If you make a mistake, cross it out clearly — do not leave the examiner in any doubt as to which material you wish to be marked. Then continue with what you believe is the correct solution.
- Look at the answer: is it sensible? A calculation that gives the mass of the Moon as 1.73×10^{4} kg is clearly wrong. If you think the answer is wrong:
  - check that you have copied the values down correctly from the question paper
  - do the calculation a second time in case you made a mistake using your calculator
  - check that you have not made a mistake in the powers of 10, a common error
- Don't forget the units at the end.

5 Using calculators

Practise using your calculator to do calculations that you know to be correct.

It may sound obvious, but make sure you can use your calculator. Do you know how to enter exponents — numbers such as 1.60×10^{-19}, the charge on an electron? A common error here is to punch in 1.60 and then use the **×** button followed by 10 **–** 19. All calculators have an exponent button, usually labelled **EE** or **Exp**, so the key sequence to insert the charge on an electron would be 1.60 **EE** 10 **+/–** 19.

The screen of the calculator should display the following:

Do you know how to obtain the log of a number or the exponential of a number using your calculator? See Topic 2, Section 3.

Note that negative numbers are usually inserted with the change sign button labelled **+/–**. Do not use the button marked **+** or **–**.

5.1 Trigonometric calculations

When you want to find the **sin**, **cos** or **tan** of an angle, your calculator will usually accept the angle in either degrees or radians, provided the correct button or switch has been set. With modern liquid crystal displays, this is usually indicated by text on the display, usually **rad** for radians or **deg** for degrees.

With some older calculators you have to use two buttons: **ARC** followed by **SIN**.

Remember, the reverse process, the conversion of, say, a sine value into an angle, uses the button marked **sin^{-1}**. The answer will be in degrees or radians, depending on which mode the calculator is set to.

6 Graphs

You may be asked to read values from a graph. In the practical examination, you will certainly be asked to plot graphs and use them to obtain a numerical result. A clear understanding of graphs is therefore vital.

6.1 Defining the scale of an axis

The units for the quantity plotted on an axis are specified as follows. Say a graph is used to plot velocity, measured in m s^{-1}, against time, measured in seconds. The modern convention used to indicate the units is as shown.

Velocity/m s^{-1}

Time/s^{-1}

The units of the quantity plotted on the axis are placed after the slash.

6.2 Straight-line graphs through the origin

The simplest graph is a straight-line graph that passes through the origin.

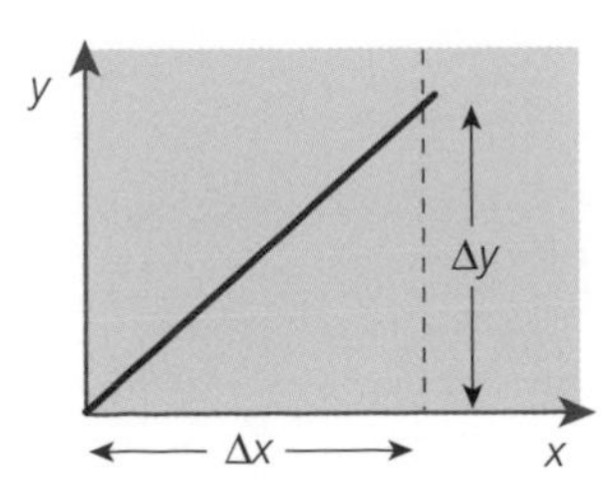

The equation of the graph is:

$$y = mx$$

where m is the slope of the graph, $\frac{\Delta y}{\Delta x}$.

6.3 Straight-line graphs not through the origin

Graphs do not always pass through the origin.

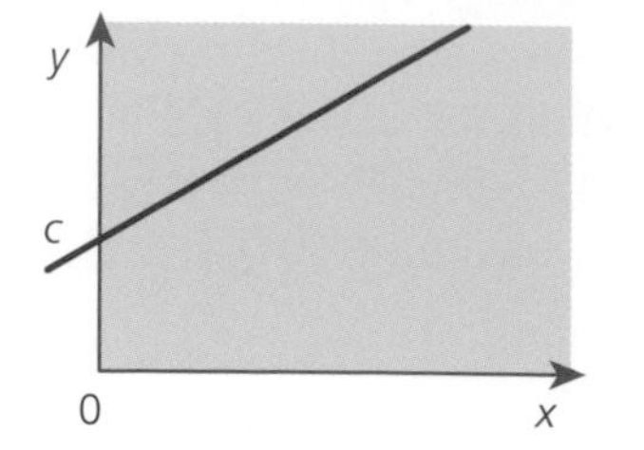

The equation of the graph is:

$$y = mx + c$$

where m is the slope of the graph and c is the value of y when $x = 0$.

6.4 Straight-line graphs not through the origin with a negative slope

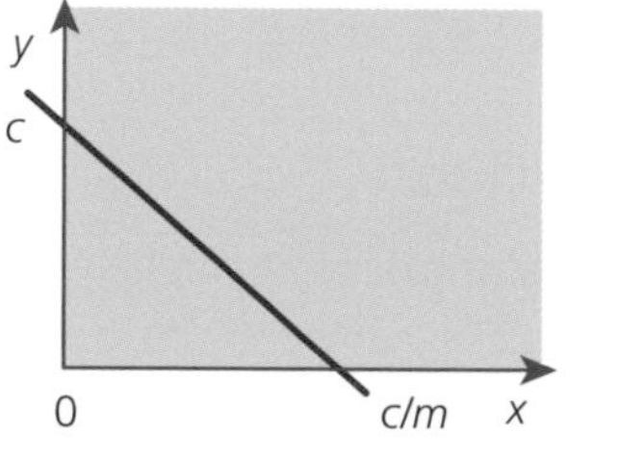

The equation of this graph is:

$$y = -mx + c$$

where $-m$ is the slope of the graph and when $x = 0$, $y = c$, or when $y = 0$, $x = c/m$.

6.5 Plotting graphs

1 Choose a scale to fill as much of the graph paper as possible. You should always choose a scale such that the plotted points fill as much of the graph paper supplied as possible, consistent with point 2 below. Your graph does not need to start with the x-axis or y-axis at the origin.

Say you were required to plot the following set of points:

x	1	2	3	4	5
y	200	210	220	230	240

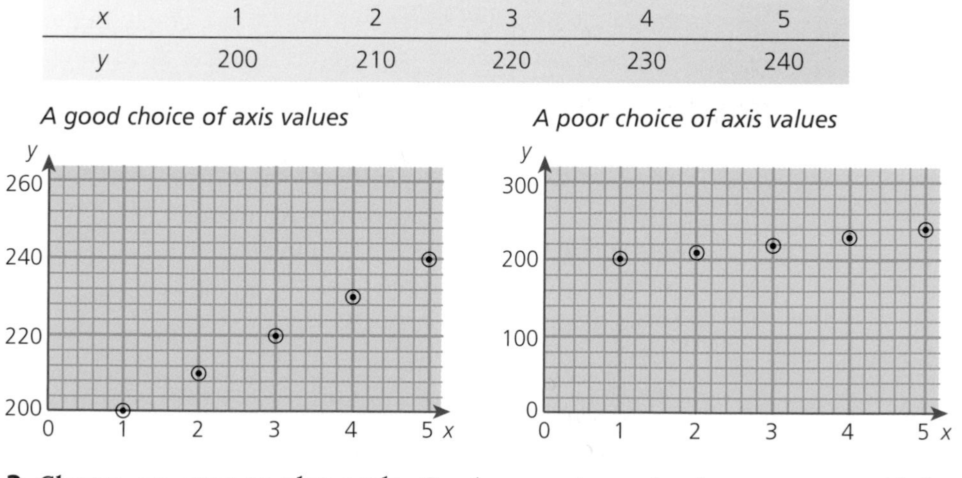

2 Choose an easy-to-plot scale. Graph paper is nearly always set out with heavy, solid grid lines subdivided into 10 units. When choosing a scale, make sure you use a scale in which it is easy to plot points.

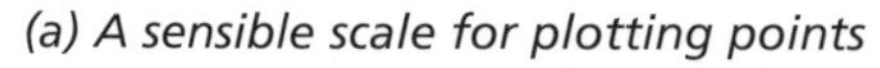

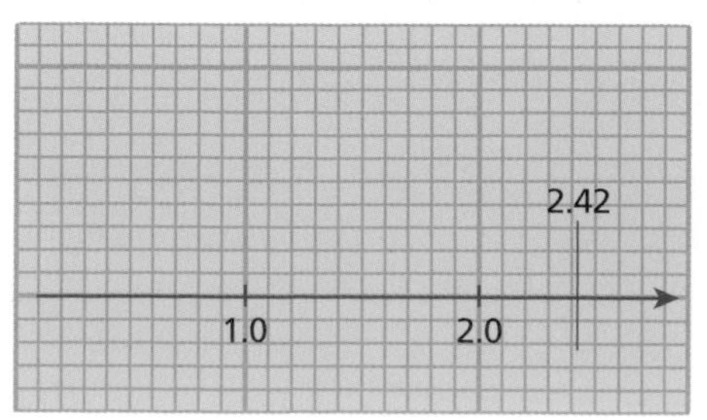

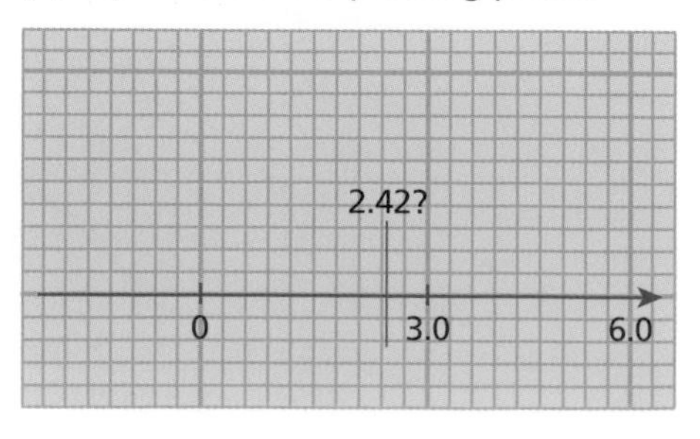

In example (a), the small subdivisions are 0.1 and locating a value such as 2.42 would be easy. In example (b), the small subdivisions are 0.3 and locating a value such as 2.42 on the x-axis would be much more difficult.

3 Finding the slope of a graph. You should use as much of the graph as possible, because this will increase the accuracy of your final result.

Do not simply draw any large triangle. Try to choose a triangle so that the values on at least one of the axes are easy to read.

This will give an accurate result

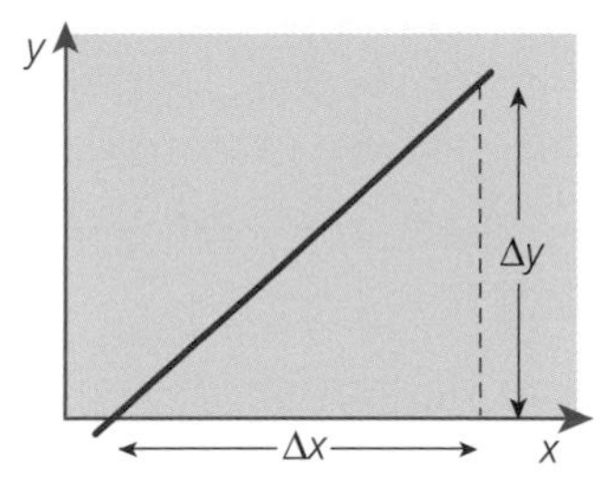

This may give an inaccurate result

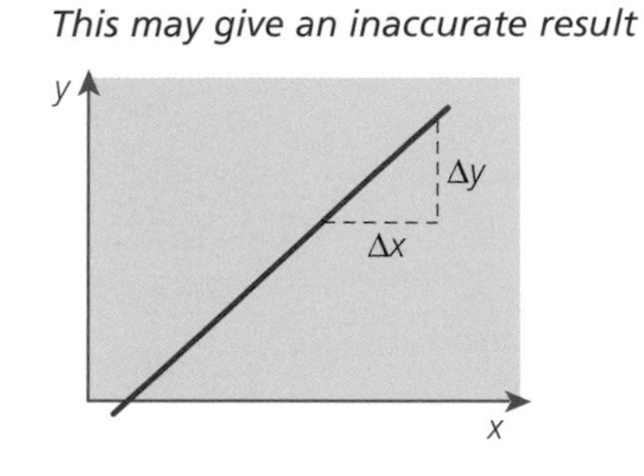

7 Drawing diagrams

In most examinations you will be asked to draw diagrams of the apparatus of an experiment or to explain some physical concept. These diagrams will nearly always be freehand sketches but you may find it convenient to use a ruler for some parts.

Some simple rules:

- The diagrams do not need to be artistic or accurate technical representations so do not spend a long time on the drawing. A simple line diagram is all that is required, but make it clear to the examiner what your sketch represents by using a label with an arrow if required. The diagram should be in proportion.

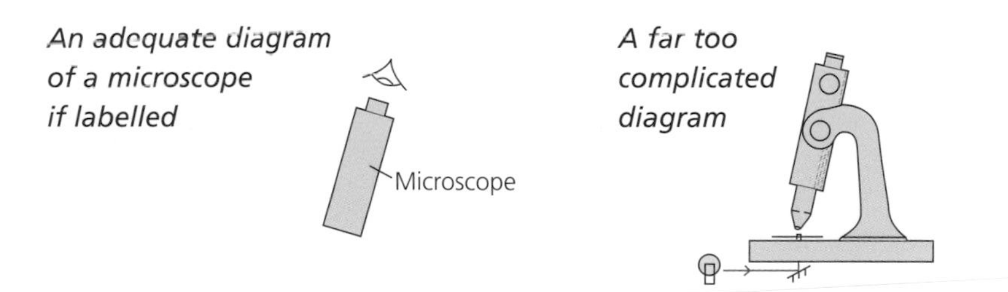

An adequate diagram of a microscope if labelled

A far too complicated diagram

- When drawing ray diagrams or indicating angles make sure they look reasonably correct.

Make sure right angles look like right angles or indicate them with a suitable label.

A good diagram showing light reflection

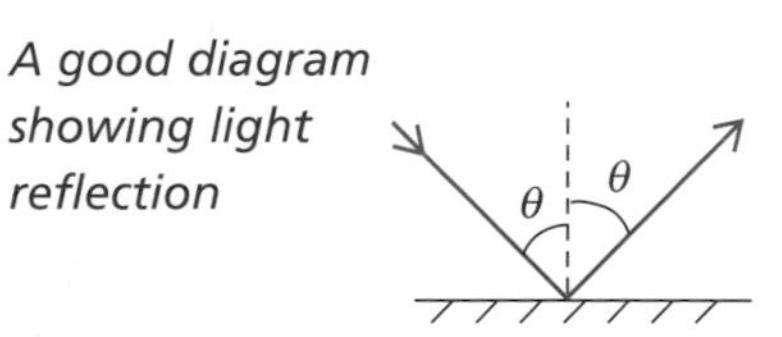

A poor diagram of light reflection

8 Examination technique and time management

The following are a few simple rules to apply when tackling examination papers to gain maximum marks:

- Prior to sitting an examination paper, make sure you know the format of the paper, how many questions are set and the time allowed.
- Calculate the approximate time you can spend on each question. Note that some questions may be awarded more marks than others and may require a longer time to give a full answer.

Don't leave questions unanswered due to lack of time.

- Do not spend too long on any one question; keep an eye on your watch.
- If you are taking too long to answer a question, leave it and continue with the rest of the paper.
- If you have time at the end, go back to the question you have not completed.
- If you have time at the end of the paper, go back and check your calculations.

9 Accepted terms used in examination questions

It is important to define the symbols you are using if you just write down the equation.

In examination papers, terms are used to indicate the type of answer required. Familiarity with these terms can help you gain maximum marks.

Define means write down a concise statement in words. Sometimes a suitable equation *with the terms defined* will suffice.

Explain is an extension of define. Having provided an initial definition, you are expected to add some additional comments.

State requires a concise answer, with little or no additional comment, and is often derived from the result of some calculation.

Describe requires a fuller answer. In many cases, an appropriate diagram will enhance the description.

Discuss requires you to look critically at the points raised in the question and make some sensible comment.

Deduce/predict requires you to make use of the information given, or additional information obtained as a result of a calculation, to answer the question.

Suggest usually means that there may be no unique answer, or that the answer may be something you have not covered on the specification, and you are expected to draw conclusions based on your knowledge.

Calculate is used when a numerical answer is required. The answer should always include units, if appropriate, and should be given to the correct number of significant figures.

Measure means that the value required must be obtained by using a suitable instrument, a ruler for example.

Determine usually means that the value required cannot be obtained directly, but can be obtained by using some measured value and applying a suitable equation.

Show is usually used when algebraic manipulation is required.

Estimate is used when the answer required need only be given to an order of magnitude.

Sketch requires a diagram or graph that need not be absolutely accurate and can often be drawn freehand or with the aid of a ruler. The shape of the graph should, however, show the correct trend. For example, an exponential graph illustrating radioactive decay should not cut the axis. In a diagram of apparatus, the proportions should be reasonable.

TOPIC 1 Units and dimensions

The **International System of Units**, the **SI** system, uses the following basic units.

1 Unit of length

The unit of length, the **metre**, was originally based on the distance between two scratch marks on a standard metre bar held at the Bureau International des Poids et Mésures in France. This was superseded by a more precise measure, based on the wavelength of a suitable line emitted by an excited krypton atom. Today, the definition is based on the speed of light in a vacuum:

Light travels 1 m in $\frac{1}{299\,792\,458}$ seconds.

1.1 Sub-units

kilometre	$1\,\text{km} = 1 \times 10^{3}\,\text{m}$	micrometre	$1\,\mu\text{m} = 1 \times 10^{-6}\,\text{m}$
centimetre	$1\,\text{cm} = 1 \times 10^{-2}\,\text{m}$	nanometre	$1\,\text{nm} = 1 \times 10^{-9}\,\text{m}$
millimetre	$1\,\text{mm} = 1 \times 10^{-3}\,\text{m}$	picometre	$1\,\text{pm} = 1 \times 10^{-12}\,\text{m}$

2 Unit of time

The unit of time, the **second**, was originally defined as $\frac{1}{86\,400}$ of the mean solar day as the Sun passes over the meridian at Greenwich. Owing to fluctuations in the Earth's rotation and the fact that it is slowing down, the standard is now based on the vibration of atoms of caesium and is defined as the time for 9 192 631 770 vibrations.

Sub-units of time have the same notation as sub-units of length — for example, microsecond and nanosecond.

3 Unit of mass

The unit of mass, the **kilogram**, is a cylinder of platinum–iridium alloy, again held at the Bureau International des Poids et Mésures in France. As yet, there is no atomic sub-standard since, in terms of the mass of a given atom, simple weighing can be undertaken to a very high degree of accuracy.

One unit often used is the metric ton, tonne = 1×10^{3} kg. It is the metric equivalent of 1 ton.

The kilogram is 1×10^{3} g, so the sub-units are all in grams, for example, micrograms and milligrams.

3.1 Atomic mass unit

When dealing with the masses of atoms, a sub-unit called the **atomic mass unit** is used. This is based on the mass of the most abundant isotope of carbon, carbon-12.

$$1\,\text{u} = \frac{\text{mass of carbon-12}}{12} = 1.660\,56 \times 10^{-27}\,\text{kg}$$

4 Additional basic units

These units are defined in terms of particular theoretical or experimental configurations.

In addition to the three basic units above, we also have units for:

- electrical current, the **ampere**
- temperature, the **kelvin**
- luminous intensity, the **candela**
- amount of a substance in terms of a number of atoms/molecules, the **mole**

- angle, the **radian**
- solid angle, the **steradian**

5 Derived units

Many of the physical quantities we measure have their own names, but they are all derived from combinations of base units.

These are just a few from a long list.

Physical quantity	Name	Symbol	Base units
Energy	joule	J	kg m^2 s^{-2}
Electrical charge	coulomb	C	A s
Pressure	pascal	Pa	N m^{-2} = kg m^{-1} s^{-2}

5.1 Other derived units

Scientists find it useful to give particular values of some quantities their own name, so that they become derived units.

Physical quantity	Name	Symbol	Magnitude
Energy	electronvolt	eV	1.60×10^{-19} J
Pressure	bar	bar	1×10^5 N m^{-2}

6 Dimensions

The area of an object is found by multiplying a length by a length. Whether we measure the length in metres, feet or inches, area has the dimensions of length squared:

$$[\text{area}] = [L]^2 \text{ where } [L] = [\text{length}]$$

Note the use of square brackets when we refer to the **dimensions** of a quantity.

Other examples are:

$$[\text{acceleration}] = \frac{\text{velocity}}{\text{time}} = [LT^{-1}\ T^{-1}] = [LT^{-2}] \text{ where } [T] = [\text{time}]$$

$$[\text{pressure}] = \frac{\text{force}}{\text{area}} = [ML^{-1}T^{-2} \text{ where } [M] = [\text{mass}]$$

Often in examinations students write dimensions incorrectly in terms of units kg, m and s. Dimensions using M, L and T are independent of units.

6.1 Dimensional analysis

One use of dimensions is to check for incorrect equations. If the dimensions on both sides of the equation are not the same, the equation cannot be correct.

An equation may be dimensionally correct but that is no guarantee that the formula is physically correct. The equation may require a constant term or the wrong variables may have been combined.

Example

The escape velocity of a satellite from the surface of the Earth is given by the equation $v = \sqrt{2gR}$, where g is an acceleration and R is the radius of the Earth.

We can check this equation using dimensional analysis:

$$\text{velocity} = [LT^{-1}] \text{ and } \sqrt{2gR} = \sqrt{[LT^{-2}][L]} = [LT^{-1}]$$

The two sides of the equation agree, so the equation is dimensionally correct.

Worked examples

Q1 What are the dimensions of momentum, density and angle?

momentum = mass × velocity = $[M][LT^{-1}] = [MLT^{-1}]$

$$\text{density} = \frac{\text{mass}}{\text{volume}} = \frac{[M]}{[L^3]} = [ML^{-3}]$$

$$\text{angle} = \theta = \frac{\text{arc length}}{\text{radius}} = \frac{[L]}{[L]} = \text{no dimensions}$$

Q2 It is suggested that kinetic energy = $\frac{1}{2}mat^2$, where m is mass, a is acceleration and t is time. Could this be a correct equation?

The correct dimensions of kinetic energy = $\frac{1}{2}mv^2 = [M][LT^{-1}]^2 = [ML^2T^{-2}]$

Using $\frac{1}{2}mat^2 = [M][LT^{-2}][T^2] = [ML^2]$

Hence, $\frac{1}{2}mat^2$ is not a correct equation for kinetic energy.

Q3 Newton's law of gravitation states: $F = G\frac{m^1m^2}{d^2}$

where F is force, m is mass and d is distance. What are the dimensions of G?

force = mass × acceleration = $[M][LT^{-2}]$

Rearranging Newton's law:

$$G = \frac{Fd^2}{m^1m^2} = \frac{[M][LT^{-2}][L^2]}{[M^2]}$$

$$= [M^{-1}L^3T^{-2}]$$

You should now know:

- **the base units in the SI system**
- **derived units**
- **the use of dimensions and dimensional analysis to verify equations**

TOPIC 2

Basic mathematics

A complete understanding of physics at A-level requires the support of some elements of mathematics. You need to be confident about the manipulation of numbers, arithmetic and the use of equations, as these will help you to understand the relationship between physical quantities. This section is intended as a quick revision of the subject. If there are any areas that cause difficulty, you should refer to a suitable teaching text.

Calculators have made much of the tedious arithmetic easier; however, you need to be proficient in their operation to get correct results. This topic works through some arithmetic calculations and explains functions such as e^x, log x, ln x, and trigonometric functions (sines, cosines and tangents) and their inverse. Try performing the calculations on your calculator to see whether you get the same answers.

1 Powers of ten

In physics we deal with very large and very small numbers, which, for convenience, are expressed in powers of ten.

10^9	1 000 000 000
10^6	1 000 000
10^3	1000
10^1	10
10^{-3}	0.001
10^{-6}	0.000 001
10^{-9}	0.000 000 001

Multiplying powers of 10

$$10^6 \times 10^3 = 10^{6+3} = 10^9$$

$$10^4 \times 10^{-2} = 10^{4-2} = 10^2$$

Dividing powers of ten

$$\frac{10^6 \times 10^4}{10^2} = 10^{6+4-2} = 10^8$$

$$\frac{10}{10^2 \times 10^{14}} = 10^{1-2-14} = 10^{-15}$$

Numbers in general

$$2000 = 2 \times 10^3$$

$$3\,000\,000 = 3 \times 10^6$$

$$16\,700 = 1.67 \times 10^4$$

Powers in general

$$2 \times 2 = 2^{1+1} = 2^2$$

$$2^2 \times 2^4 = 2^{2+4} = 2^6$$

2 Logarithms

Logarithms are a convenient way of dealing with numbers as they often simplify the mathematics. Some physical processes obey a logarithmic relationship.

If $y = a^x$, then x is said to be the logarithm of y to the base a, written as $x = \log_a y$.

If $y_1 = a^j$ and $y_2 = a^k$, then:

$$y_1 y_2 = a^j a^k = a^{j+k}$$

So:

$$\log_a y_1 y_2 = j + k = \log_a y_1 + \log_a y_2$$

$$\log_a \frac{y_1}{y_2} = j - k = \log_a y_1 - \log_a y_2$$

Also:

$$\log_a y^n = n\log_a y$$

$$\log_a a = 1$$

$$\log_a 1 = 0$$

2.1 Common and natural logarithms

Two bases are used in physics: base 10, or common logarithms, and base e, or natural logarithms, where e = 2.728… .

2.2 Common logarithms

$\log_{10} 10 = 1$ since $10 = 10^1$

$\log_{10} 1000 = 3$

$\log_{10} 0.001 = -3$

The above are easy to work out but the logs of numbers that are not powers of ten require the use of the **log** key on your calculator.

It is usual just to write log 57.3, in which case base 10 is assumed.

Example

log 57.3 = 1.758 and log 0.00078 = −3.108

2.3 Natural logarithms

In natural logarithms, when $y = e^x$, where e = 2.728, then $y = \ln_e x$.

Use the **ln** key on the calculator for natural logarithms.

Example

ln 57.3 = 4.048 and ln 0.00078 = −7.156

2.4 Inverse of logarithms

To convert back to numbers from logarithms use the **10^x** button on the calculator for base 10 and the **e^x** button for natural logarithms.

Example

$\log^{-1} 6.345 = 2.213 \times 10^6$ and $\ln^{-1} -4.285 = 0.0138$

3 *The exponential function*

Many physical systems obey exponential functions, which occur when the rate of change of a quantity is proportional to the quantity itself. Radioactive decay is one such example.

Note: when $x = 0$, e^x and $e^{-x} = 1$.

3.1 Exponential curves

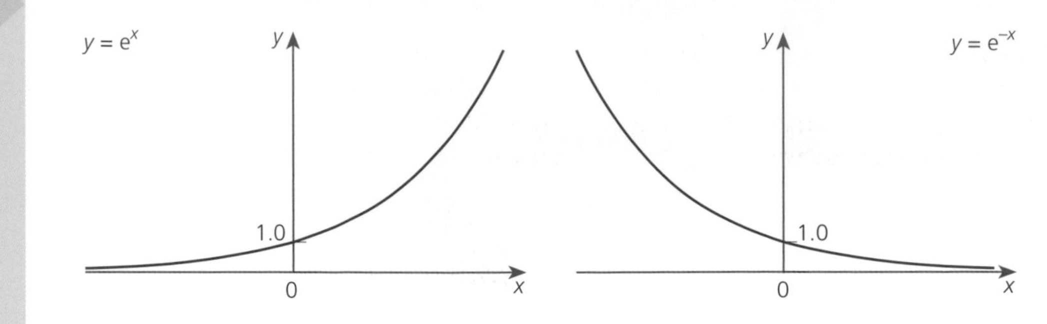

4 The binomial expansion

The binomial expansion is useful for making approximations, e.g. $(1 + x)^n = 1 + nx$ provided x is small, $\ll 1$.

5 Quadratic equations

Given the quadratic equation:

$$ax^2 + bx + c = 0$$

the two solutions are:

$$x = \frac{-b + \sqrt{b^2 - 4ac}}{2a} \text{ and } \frac{-b - \sqrt{b^2 - 4ac}}{2a}$$

6 Rearranging equations

When rearranging equations, you are allowed to add, subtract, divide or multiply the equation by anything, provided you do the same on both sides of the equal sign. Look at the following examples.

(i) $A = \pi r^2$

Rearrange to make r the subject:

$$\frac{A}{\pi} = \frac{\cancel{\pi}}{\cancel{\pi}}r^2$$

$$\sqrt{\frac{A}{\pi}} = \sqrt{r^2}$$

$$\sqrt{\frac{A}{\pi}} = r$$

(ii) $m_1 - m_2 = -2.5\log\left(\frac{b_1}{b_2}\right)$

Rearrange to make b_1 the subject:

$$\frac{m_1 - m_2}{-2.5} = \frac{\cancel{-2.5}}{\cancel{-2.5}}\log\left(\frac{b_1}{b_2}\right)$$

$$10^{-((m_1 - m_2)/2.5)} = \frac{b_1}{b_2}$$

$$b_2 10^{-((m_1 - m_2)/2.5)} = b_1$$

7 Order of evaluation

In any calculation, the order in which the operations are performed is important. A good rule to follow is **BODMAS** (**B**rackets, **O**f, **D**ivision, **M**ultiplication, **A**ddition and **S**ubtraction), which means we must first work out the brackets in an equation, then any functions, then division and multiplication, followed by addition and subtraction.

Examples

Some simple calculations using a calculator:

- Given $L = 4\pi r^2 \sigma T^4$, evaluate L when $T = 6000$, $r = 6.0 \times 10^8$ and $\sigma = 5.67 \times 10^{-8}$.

  $L = 4\pi(6.0 \times 10^8)^2 \times 5.67 \times 10^{-8} \times (6000)^4$

  $L = 3.3 \times 10^{26}$

- Given $M = \frac{4}{3}\pi r^3 \rho$, evaluate M when $r = 7 \times 10^8$ and $\rho = 1400$.

  $M = \frac{4}{3} \times \pi \times (7 \times 10^8)^3 \times 1400$

  $M = 2 \times 10^{30}$

- Given $m_1 - m_2 = -2.5 \times \log\left(\frac{b_1}{b_2}\right)$, evaluate m_1 when $m_2 = 3.0$, $b_1 = 120$ and $b_2 = 30$.

  $m_1 = 3 - 2.5 \times \log\left(\frac{120}{30}\right)$

  $m_1 = 1.5$

In this calculation, remember BODMAS and evaluate the brackets first, then the log, then multiply, and finally do the subtraction.

8 Geometry

The ratio of the circumference of a circle to its diameter is a constant, $\pi = 3.141592\ldots$

circumference $= \pi d = 2\pi r$

area of a circle $= \pi r^2$

volume of a cylinder $= \pi r^2 L$

surface area of a sphere $= 4\pi r^2$

volume of a sphere $= \frac{4}{3}\pi r^3$

9 Trigonometry

9.1 Degrees and radians

The angular separation between two lines can be measured in degrees or radians.

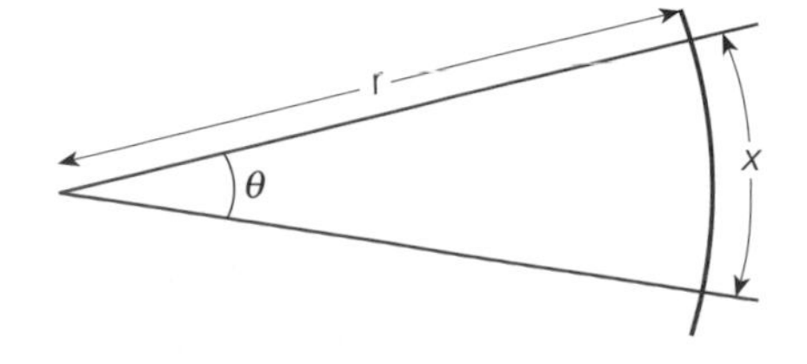

The angle θ is related to x and r by the equation $\theta = \frac{x}{r}$, and in a full circle there are 2π radians. We can convert between angles in radians and degrees as follows:

1 radian is the angle when $x = r$.

$$\text{radians} = \frac{2\pi}{360} \times \text{degrees} \qquad \text{degrees} = \frac{360}{2\pi} \times \text{radians}$$

Examples

30° = 0.524 radians 3.4 radians = 195°

Here are some useful relationships between angles:

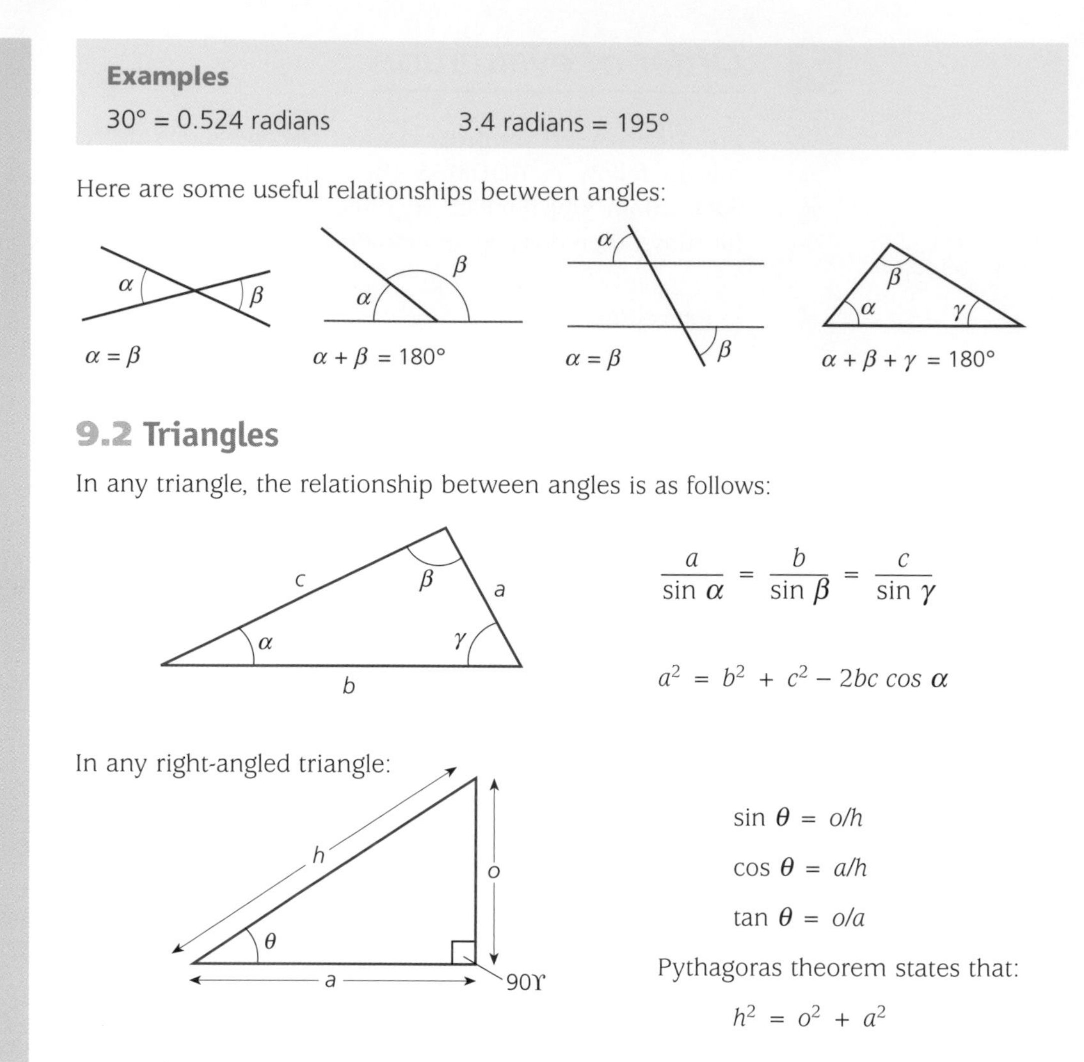

9.2 Triangles

In any triangle, the relationship between angles is as follows:

$$\frac{a}{\sin \alpha} = \frac{b}{\sin \beta} = \frac{c}{\sin \gamma}$$

$$a^2 = b^2 + c^2 - 2bc \cos \alpha$$

In any right-angled triangle:

$\sin \theta = o/h$

$\cos \theta = a/h$

$\tan \theta = o/a$

Pythagoras theorem states that:

$h^2 = o^2 + a^2$

9.3 Sine, cosine and tangent curves

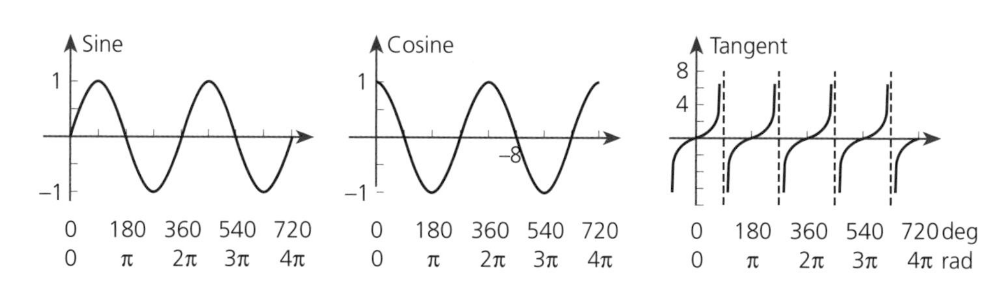

10 Using the calculator

Since angles measured in either degrees or radians can obtain correct results for the sin, cos or the tan of an angle, you need to be sure which unit you are using and that the correct switch is set on your calculator.

The sine of 45° = 0.707 The sine of 1.4 radians = 0.985

Given the sine (or cosine or tangent) of an angle, the reverse process is to find the angle in either degrees or radians (depending on which you are using) by pressing the $\sin^{-1}$ (or $\cos^{-1}$ or $\tan^{-1}$) button on the calculator.

$\sin^{-1} 0.255 = 14.8°$ or 0.258 radians

11 Calculus

Differentiation gives the gradient of a graph.

If $y = x^n$ then the gradient $\frac{dy}{dx}$ is given by $\frac{dy}{dx} = nx^{n-1}$

If $y = \sin x$ $\qquad \frac{dy}{dx} = \cos x$

If $y = \cos x$ $\qquad \frac{dy}{dx} = -\sin x$

If $y = e^x$ $\qquad \frac{dy}{dx} = e^x$

Integration is the reverse of differentiation and gives the area under a curve.

If $y = x^n$ then the integral, $\int y\,dx$ is given by $\int y\,dx = \frac{x^{n+1}}{n+1} + c$

If $y = \sin x$ $\qquad \int y\,dx = -\cos x + c$

If $y = \cos x$ $\qquad \int y\,dx = \sin x + c$

If $y = \frac{1}{x}$ $\qquad \int y\,dx = \ln x + c$

If $y = e^x$ $\qquad \int y\,dx = e^x + c$

where c is a constant.

TOPIC 3 Mechanics

1 Scalars and vectors

A **scalar** quantity has only magnitude, for example *distance, mass, temperature, speed.*

A **vector** quantity has both magnitude and direction, for example *displacement, weight, force, velocity.*

For some quantities, different words are used to differentiate between scalar and vector values:

Scalar	Vector
distance	displacement
speed	velocity
mass	weight

1.1 Graphical representation of vectors

A vector can be represented by a line of suitable length drawn in the direction in which it acts.

Example

A force of 4 newtons acting to the left can be represented by a line as follows. Let 1 cm = 1 newton, then the force can be represented by a line 4 cm long, pointing in the correct direction.

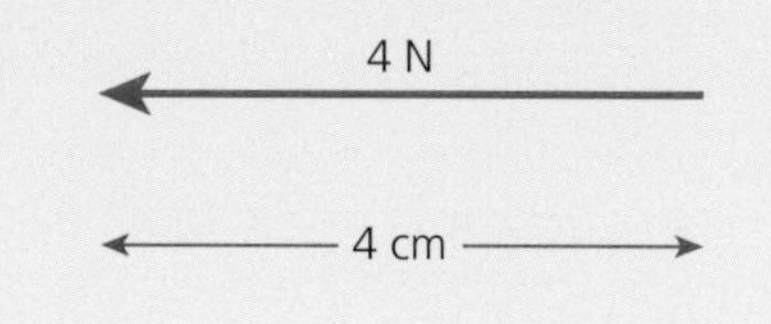

Always choose an easy scale to draw vectors. For an accurate result, make the diagram as large as possible.

Vector quantities in equations are often represented in bold type, **A**, or with an arrow drawn above, $\overrightarrow{AB}$.

1.2 Addition of scalars and vectors

To add **scalar** quantities together you simply add the numerical values together.

To add **vector** quantities together you must consider both the magnitude and the direction of the vectors in the addition.

Addition of two vector quantities

To add two vector quantities, the **head-to-tail** rule is used.

To add two vectors together, first join the tail of the second vector to the head of the first vector, each drawn with the correct magnitude and direction. The **resultant** vector is obtained in magnitude and direction by drawing a line joining the tail of the first vector to the head of the second vector.

Draw the diagram with care. Make sure the angle between the vectors is correct, as an incorrect value can lead to a large error in the answer.

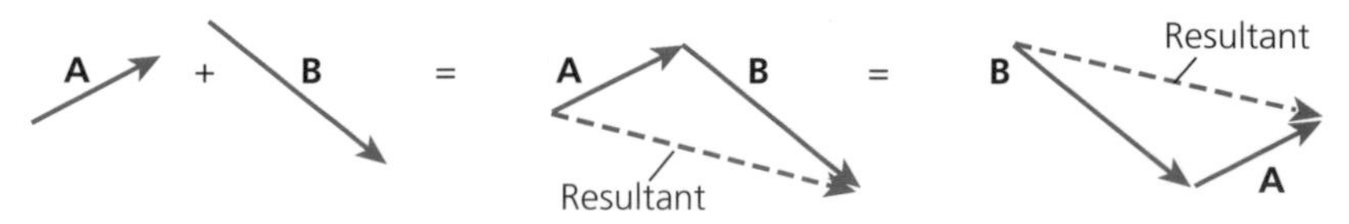

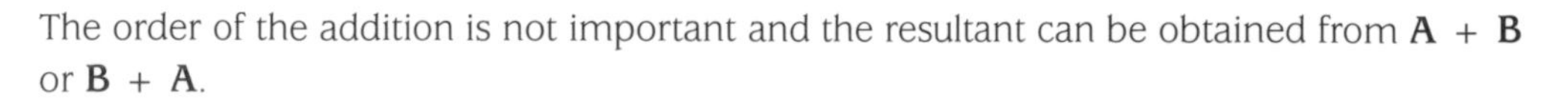

The order of the addition is not important and the resultant can be obtained from **A** + **B** or **B** + **A**.

Addition of two vectors at right angles

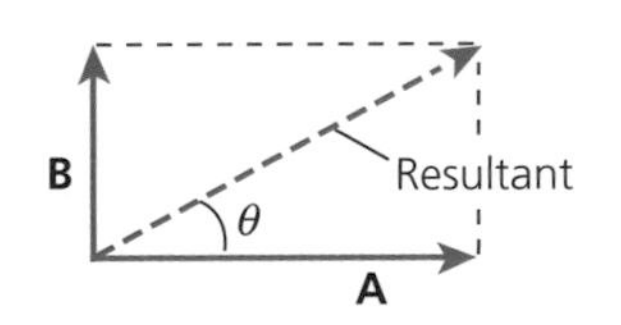

Magnitude of the resultant vector $= \sqrt{A^2 + B^2}$

The direction is given by: $\tan \theta = \dfrac{B}{A}$

1.3 Resolution of a single vector into two at right angles

There are many occasions in physics where it is convenient to convert a single vector into two vectors at right angles.

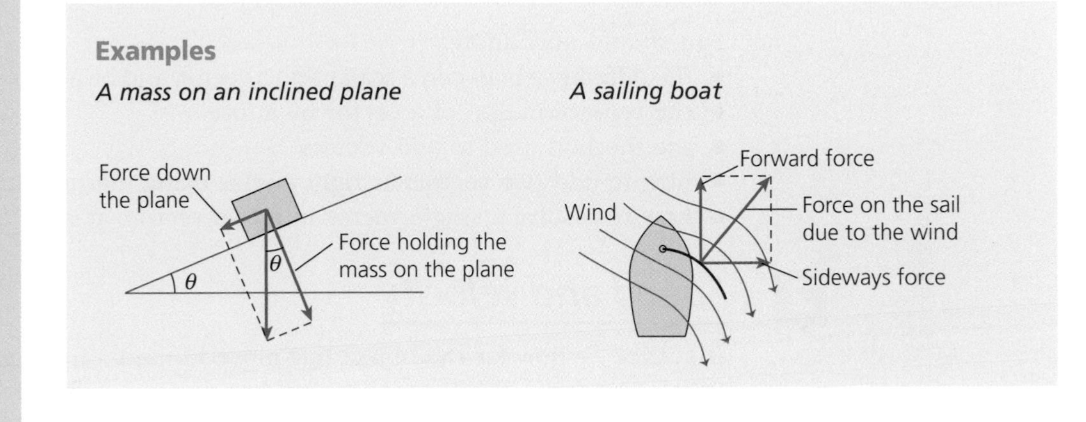

Magnitude of the two components

The following vector **A** can be resolved into two vectors as follows:

If the angle α is to be used, the equations are reversed.

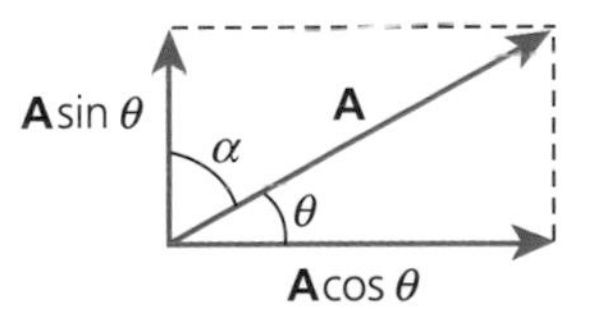

magnitude of the horizontal component $= \mathbf{A}\cos\theta$

magnitude of the vertical component $= \mathbf{A}\sin\theta$

Worked examples

Q1 What is the resultant force on the block?

2 N   5 N

Q2 A ship leaves port and sails due north for 3 km and then northwest for 5 km. What is the displacement of the ship from the port?

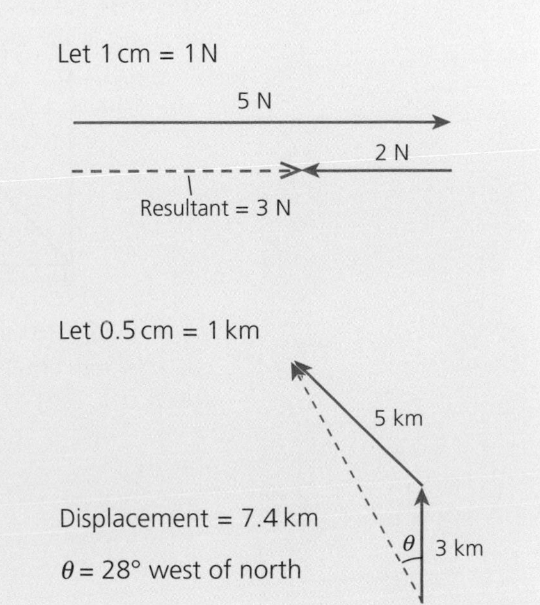

Q3 What is the vertical force exerted by the block and the angle of the slope with respect to the horizontal?

1.7 N
θ
4.7 N

resultant force $= \sqrt{4.7^2 + 1.7^2}$

$= \sqrt{25} = 5\,\text{N}$

$\tan\theta = \dfrac{1.7}{4.7} = 0.362$

$\theta = 20°$

1.7
θ
4.7

You should now know:

- the difference between a scalar and a vector, and be able to give examples of each
- the representation of a vector by a line
- the method used to add vectors
- how to add two vectors at right angles using mathematical equations
- how to resolve a single vector into two vectors at right angles

2 Speed and velocity

Distance — how far the object has moved in any direction (a **scalar** quantity).
Displacement — the distance moved in a particular direction (a **vector** quantity).
Speed — the distance moved divided by the time taken (a **scalar** quantity).
Velocity — the displacement divided by the time taken (a **vector** quantity).

In some calculations, you may only be required to calculate the numerical values of velocity and acceleration. However, in calculations involving force, the direction of the force is important and hence its effect on the velocity.

2.1 Distance–time graphs

Distance–time graphs are a convenient way to indicate how the speed of an object varies with time.

With constant speed, a straight-line graph is obtained and the speed is measured from the slope of the graph.

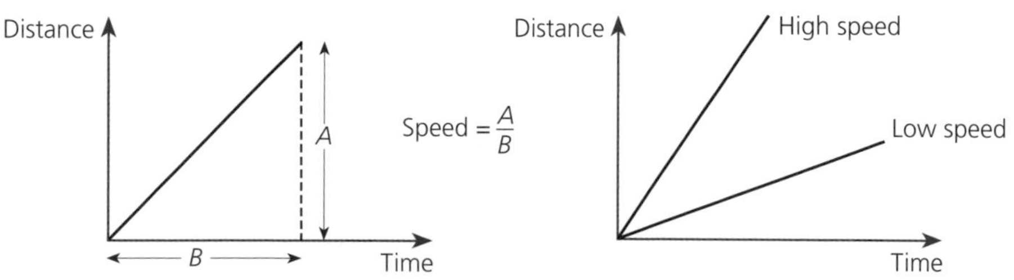

When the speed varies, the object undergoes an acceleration and the slope of the graph changes with time. Measurement of the slope at a particular time gives the instantaneous speed.

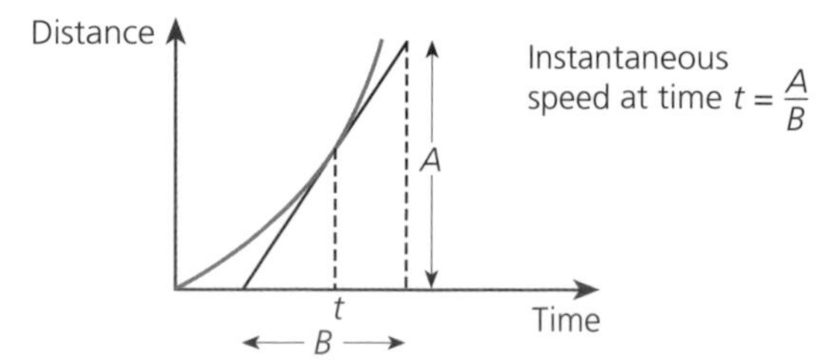

2.2 Displacement–time graphs

Sometimes it is convenient to draw a displacement–time graph. One example is the oscillation of a particle about an equilibrium position, where the slope of the graph at any instant gives both the magnitude and direction of the velocity with respect to the equilibrium.

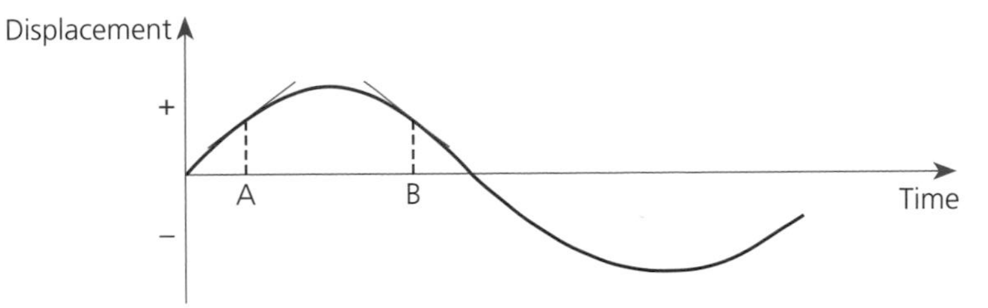

At A the slope is positive and the velocity is away from the equilibrium position. At B the slope is negative and the velocity is towards the equilibrium position.

2.3 Speed–time graphs

$$\textbf{speed} = \frac{\textbf{distance}}{\textbf{time}}$$

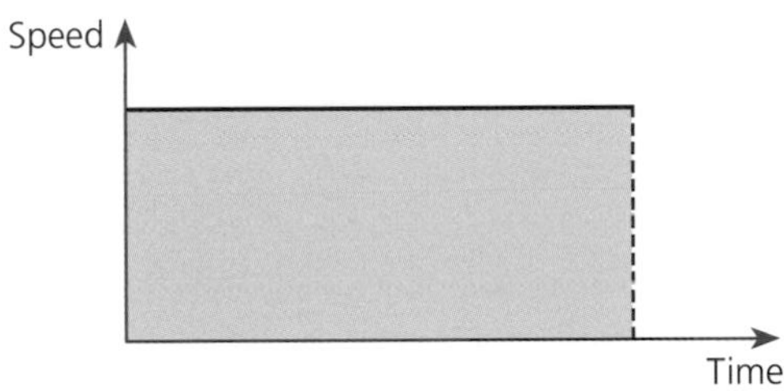

When an object moves at a constant speed, the speed–time graph is a horizontal straight line.

Rearranging the above equation gives:

$$\textbf{distance} = \textbf{speed} \times \textbf{time}$$
$$= \textbf{the area under the curve}$$

This applies to any speed–time graph, even when the speed changes with time (see Section 3.3).

Worked examples

Q1 An object travels a distance of 4 m in 10 s at a constant speed. It then stops for 4 s before travelling a further 8 m in 12 s. Draw a distance–time graph and calculate the speed during the three time intervals.

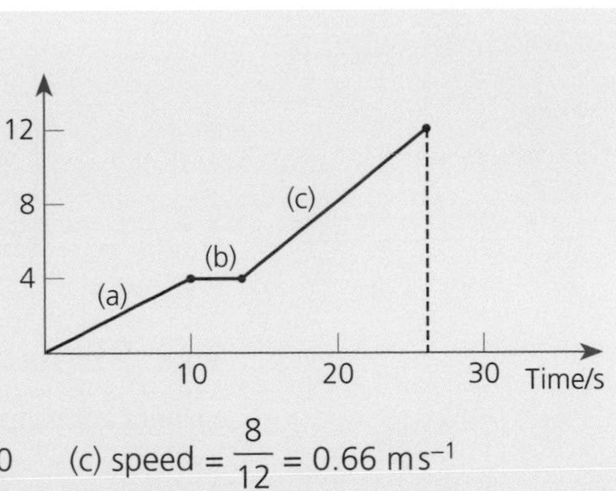

(a) speed $= \frac{4}{10} = 0.4\ \text{m s}^{-1}$ (b) speed = 0 (c) speed $= \frac{8}{12} = 0.66\ \text{m s}^{-1}$

Q2 A rubber ball is thrown at a wall and moves with a constant speed of $5\ \text{ms}^{-1}$. It then bounces off the wall with the same speed. Plot a displacement–time graph, with respect to the wall, for the motion of the ball from 10 s prior to hitting the wall to 10 s after.

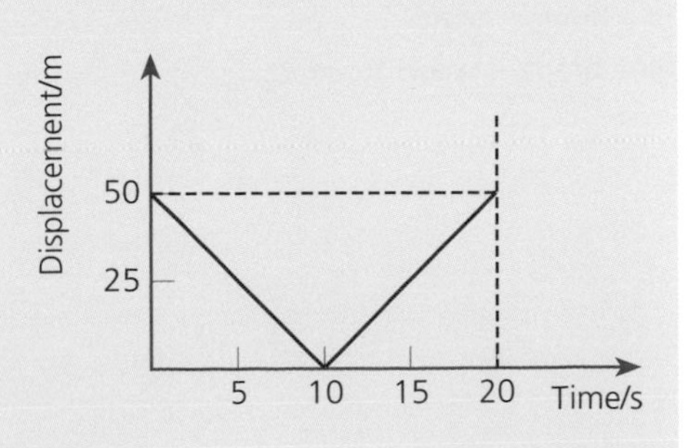

In 10 s, the ball moves a distance $5 \times 10 = 50$ m. When moving towards the wall, the velocity is negative with respect to displacement from the wall.

Q3 In the above question, if the ball leaves the wall with a speed of $3\,m\,s^{-1}$, draw a speed–time graph and calculate the total distance travelled during the 20 s.

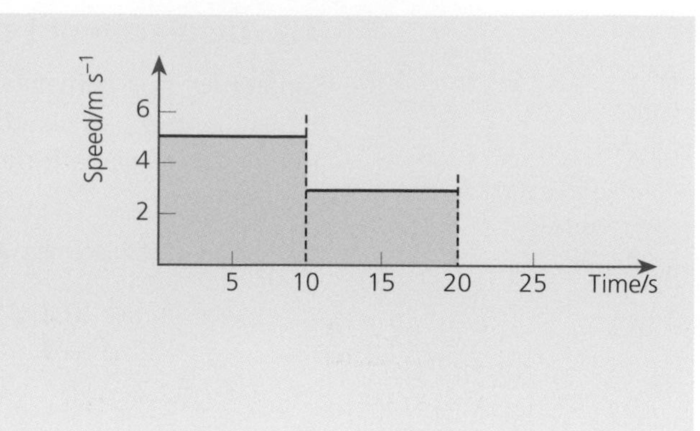

Distance travelled
= area under the graph
$= 5 \times 10 + 3 \times 10$
= 80 m

You should now know:

- **the definitions of distance, displacement, speed and velocity**
- **how to draw a distance–time graph and calculate the speed**
- **an example of a displacement–time graph and how to calculate the velocity**
- **how to calculate the distance travelled using a speed–time graph**

3 Accelerated motion graphs

3.1 Distance–time graphs

Remember, the slope of a distance–time graph is speed.

When an object is moving with a constant acceleration, the distance–time graph is a curved line, the distance moved being proportional to the square of the time when accelerating.

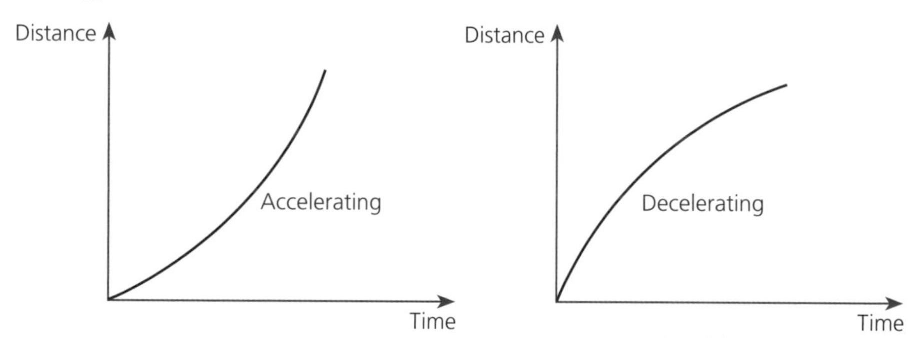

3.2 Displacement–time graphs

Remember, the slope of a displacement–time graph is velocity.

Similar rules apply to displacement–time graphs when the object moves under constant acceleration.

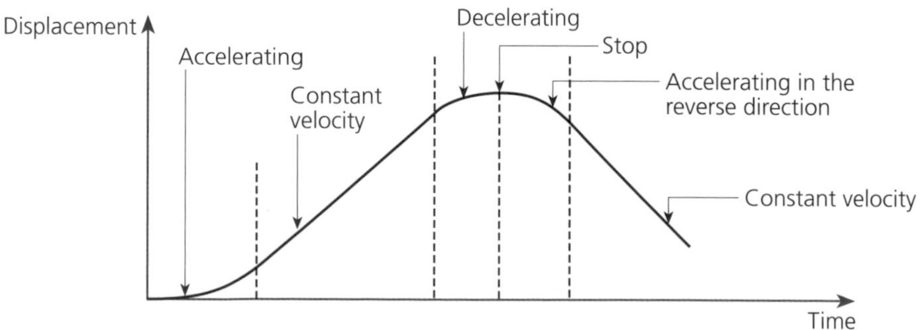

3.3 Speed–time graphs

Speed–time graphs are a convenient way of indicating how the acceleration of an object changes with time.

Note that when the acceleration changes with time, the slope at a given time t is the instantaneous value of the acceleration.

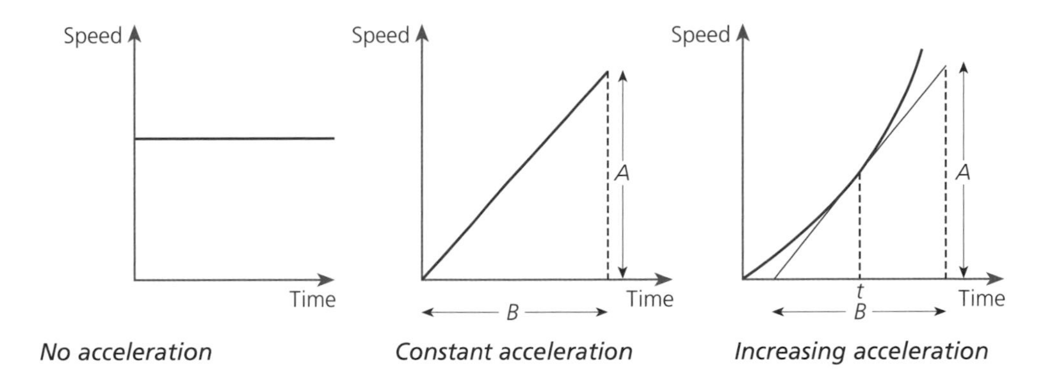

The slope of the speed–time graph is the acceleration: acceleration $= \frac{A}{B}$.

3.4 Speed–time graphs — distance travelled

Speed–time graphs are also valuable since the area under the graph is the distance travelled.

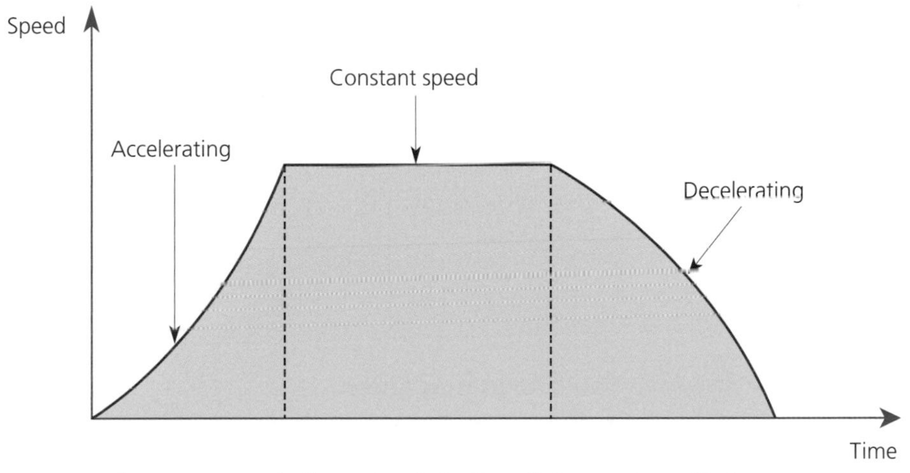

The area under any speed–time graph is the distance travelled.

3.5 Velocity–time graphs

A velocity–time graph can yield important information about the motion of a body — for example, when a body is oscillating or when the motion reverses in direction.

Worked examples

Q1 A car stars from rest and accelerates to a speed of 7 m s^{-1} in 20 s. It then travels for 40 s at this speed. Draw a speed–time graph. Calculate the initial acceleration and the total distance travelled.

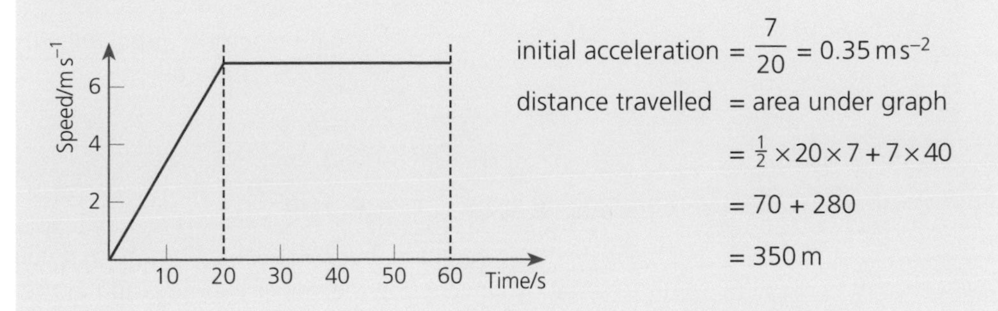

initial acceleration $= \frac{7}{20} = 0.35\,\text{m s}^{-2}$

distance travelled = area under graph

$= \frac{1}{2} \times 20 \times 7 + 7 \times 40$

$= 70 + 280$

$= 350\,\text{m}$

Q2 An object has the following velocity–time graph. Describe the motion of the object.

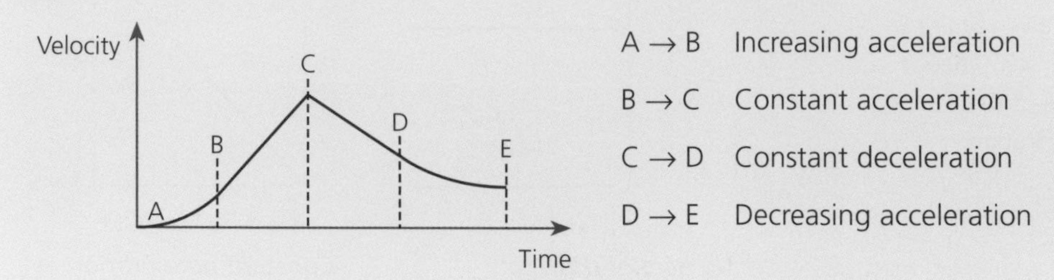

Q3 An engine shunting in a siding starts from rest and accelerates at a constant rate for 10 s. It then moves with a constant speed for 20 s, and then decelerates at a constant rate to a stop after 10 s. Next it reverses, accelerating at a faster rate for 5 s. Sketch a displacement–time graph for the motion of the train.

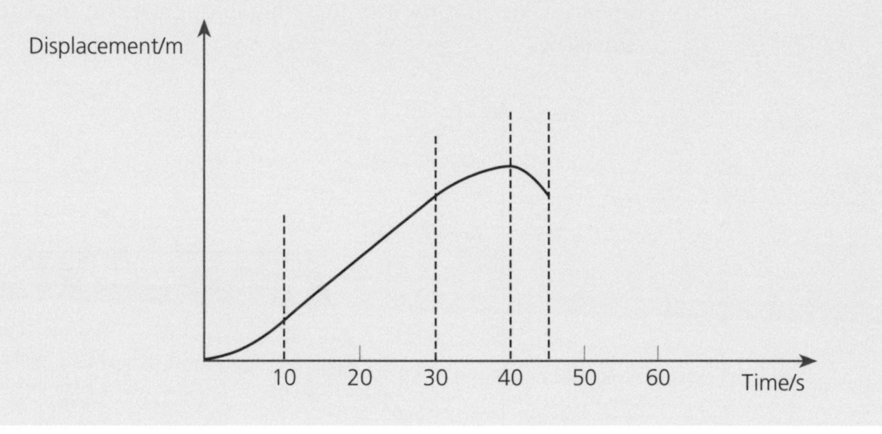

You should now know:

- **the shape of distance–time and displacement–time graphs for constant acceleration**
- **the shape of a speed–time graph for zero, constant and varying acceleration**
- **how to obtain the acceleration from the slope of a speed–time graph**
- **that the area under a speed–time graph is the distance travelled**

4 *Equations of uniformly accelerated motion*

$$\textbf{acceleration} = \frac{\textbf{change in velocity}}{\textbf{time taken}}$$

In the case of an object undergoing a constant acceleration, which starts with an initial velocity u and in a time t reaches a final velocity v, then:

$$\text{acceleration } a = \frac{\text{final velocity} - \text{initial velocity}}{\text{time taken}}$$

$$a = \frac{v - u}{t}$$

Rearranging gives: $\boldsymbol{v = u + at}$

This is the basic equation, but there are two more that you need to know. These can be understood by considering the velocity–time graph of the above object.

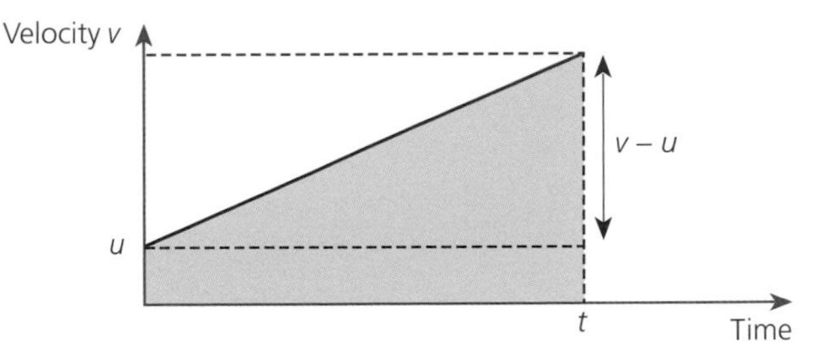

The distance travelled is the area under the graph, which is the sum of two parts — a rectangle, area $u \times t$, and a triangle, area $\frac{1}{2}(v - u)t$.

So the distance travelled is:

$$s = ut + \tfrac{1}{2}(v - u)t$$

But from our opening equation, $(v - u) = at$. So we can simplify to get:

$$\boldsymbol{s = ut + \tfrac{1}{2}at^2}$$

Eliminating t from any of the above equations gives:

$$\boldsymbol{v^2 = u^2 + 2as}$$

4.1 Average velocity

Under constant acceleration, we can also calculate the average velocity:

$$\text{average velocity} = \frac{\text{initial velocity} + \text{final velocity}}{2} = \frac{u + v}{2}$$

4.2 Deceleration

When the object is slowing down, the acceleration is negative and the three equations of motion are:

$$\boldsymbol{v = u - at} \qquad \boldsymbol{s = ut - \tfrac{1}{2}at^2} \qquad \boldsymbol{v^2 = u^2 - 2as}$$

Worked examples

Q1 A motorbike starts from rest and accelerates at $3\,\text{m s}^{-2}$. Calculate the final velocity after 15 s. If the rider then applies his brakes and comes to a stop in 6 s, what is the deceleration? What distance has the bike travelled during the braking phase of the motion?

Note that when decelerating the final velocity is zero.

Final velocity $v = u + at$ $\quad \therefore v = 0 + 3 \times 15 = 45\,\text{m s}^{-1}$

When decelerating $v = u - at$ $\quad \therefore 0 = 45 - a \times 6$

$$a = \frac{45}{6} = 7.5\,\text{m s}^{-1}$$

Using $s = ut - \frac{1}{2}at^2$ $\quad s = 45 \times 6 - \frac{1}{2} \times 7.5 \times 6^2$

$$s = 270 - 135 = 135\,\text{m}$$

Q2 Electrons in a television tube are accelerated by an electric field between two plates at a distance 3 mm apart. The electrons leave the second plate through a small hole with a velocity of $2.5 \times 10^6\,\text{m s}^{-1}$. If they start from rest at the first plate, what constant acceleration have they experienced?

Convert millimetres to metres.

$$v^2 = u^2 + 2as$$

$$(2.5 \times 10^6)^2 = 0 + 2 \times a \times 3.0 \times 10^{-3}$$

$$a = \frac{(2.5 \times 10^6)^2}{2 \times 3 \times 10^{-3}} = 1.0 \times 10^{15}\,\text{m s}^{-2}$$

Remember, always work in metres and seconds, so you will need to convert the km h^{-1}.

Q3 A holiday jet lands at a speed of 200 km h^{-1} and takes 15 s to reach its taxiing speed of 50 km h^{-1}. What length of runway must be built to ensure that only one half of the runway is used?

$$200\,\text{km h}^{-1} = \frac{200 \times 10^3}{60 \times 60} = 56\,\text{m s}^{-1} \qquad 50\,\text{km h}^{-1} = \frac{50 \times 10^3}{60 \times 60} = 13.9\,\text{m s}^{-1}$$

Calculate the deceleration using $v = u - at$:

$$13.9 = 56 - a \times 15$$

$$a = \frac{56 - 13.9}{15} = 2.8\,\text{m s}^{-2}$$

Calculate the distance travelled using $v^2 = u^2 - 2as$:

$$(13.9)^2 = 56^2 - 2 \times 2.8 \times s$$

$$s = \frac{56^2 - 13.9^2}{2 \times 2.8} = 530\,\text{m}$$

runway length = 1060 m

You should now know:

- **how to derive the three equations of motion using a velocity–time graph**
- **the form of the equations when applied to objects that are decelerating**
- **how to convert speed in km h^{-1} to speed in m s^{-1}**

5 *Gravitational acceleration*

Objects on the surface of the Earth experience gravitational forces acting towards the centre of the Earth.

When released, they accelerate downwards with a constant acceleration of 9.81 m s^{-2}, often given by the letter g (neglecting slight variations in g over the surface of the Earth and air resistance).

Similarly, an object moving upwards experiences a deceleration of the same magnitude.

The gravitational force on other planets is different. For example, on the Moon the acceleration due to gravity is about one-sixth of that on Earth.

5.1 Projectile motion

The motion of projectiles such as arrows, cricket balls, bullets etc. can be considered in two directions: along the horizontal and vertical axes.

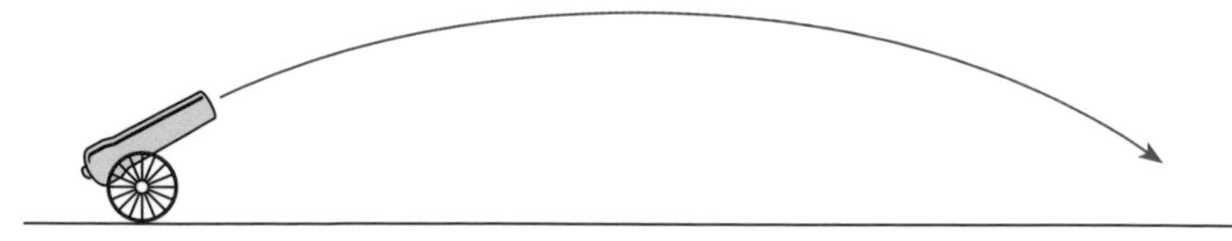

The motion along the two axes is as follows:

- in the horizontal direction the projectile moves with constant velocity
- in the vertical direction the projectile moves under the action of a gravitational acceleration

The resultant motion is a combination of the two. When performing calculations, we can deal with each motion separately.

The total time of flight is the variable linking the two parts together.

Initial horizontal and vertical velocity

In order to reach a target, projectiles are often released at an angle to the horizontal and, for performing any calculations, the initial horizontal and vertical velocities are required.

Remember, velocity is a vector quantity, and we are just resolving the velocity into two components (see Section 1.3).

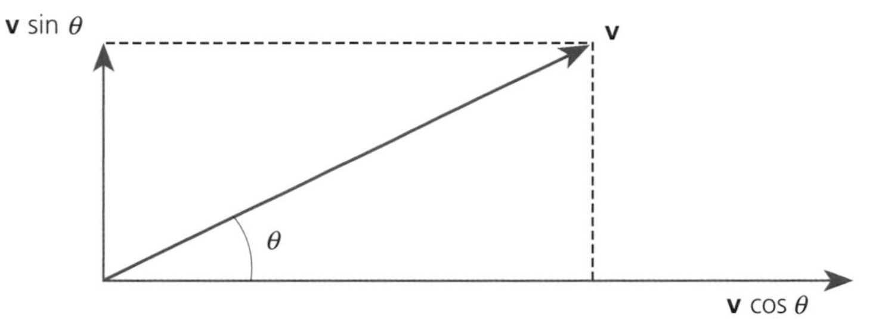

Zero initial vertical velocity

In some cases, the initial vertical velocity is zero, as with a ball thrown outwards horizontally from a cliff.

5.2 Projectile calculations

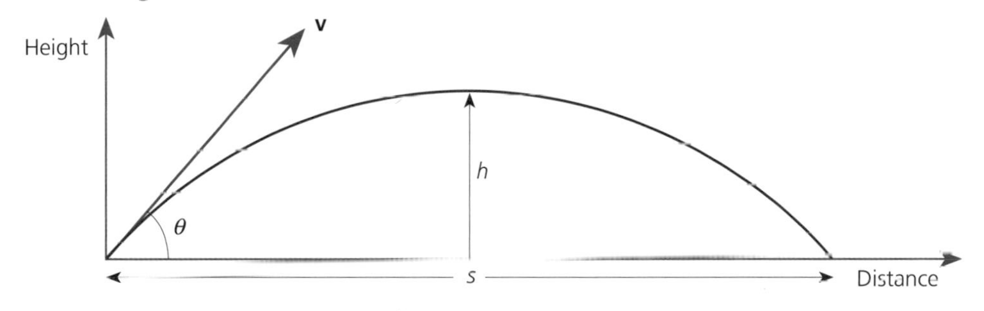

If t is the total flight time, then $\frac{t}{2}$ is the time to reach maximum height.

The only equation used with the horizontal motion is:

$$s = v \cos \theta t$$

With the vertical motion up to the maximum height, the equations are:

$$0 = v \sin \theta - g\frac{t}{2}$$

$$0^2 = (v \sin \theta)^2 - 2gh$$

$$h = v \sin \theta \frac{t}{2} - \tfrac{1}{2}g\left(\frac{t}{2}\right)^2$$

Maximum range

If we combine the above equations, the maximum range s is shown to be:

$$s = \frac{v^2 \sin 2\theta}{g}$$

It is wise not to remember this equation but to perform calculations using the four equations above.

Worked examples

Q1 A book of mass 0.3 g falls to the ground from a table of height 0.9 m. With what velocity will it strike the ground?

$$v^2 = u^2 + 2gs$$

$$v^2 = 0 + 2 \times 9.81 \times 0.9$$

$$v^2 = 17.6 \quad \therefore v = 4.2\,\text{m}\,\text{s}^{-1}$$

Note that the mass of the book is not required for the calculation.

Total time of flight = 2 × time to reach maximum height.

Q2 A ball is thrown with a velocity of 22 m s^{-1} at an angle of 18° to the horizontal. What height will the ball reach during its flight? What distance will the ball travel before it strikes the ground?

Using $v^2 = u^2 - 2gs$ $\quad 0^2 = (22 \times \sin 18)^2 - 2 \times 9.81 \times h$

$$\therefore h = \frac{(22 \times \sin 18)^2}{2 \times 9.81} = \frac{46.2}{19.6} = 2.4\,\text{m s}^{-1}$$

To calculate the time to reach maximum height, use $v = u - gt$:

$$0 = 22 \times \sin 18 - 9.81 \times t \qquad t = \frac{22 \times \sin 18}{9.81} = 0.69\,\text{s}$$

Considering the horizontal motion $s = v \times 2 \times t = 22 \times \cos 18 \times 2 \times 0.69 = 29$ m

Q3 A gun is positioned in a cliff 80 m above sea level. It fires a shell, which leaves the muzzle with a velocity of 400 m s^{-1} in a horizontal direction. What is the closest distance that a ship can approach the face of the cliff without being hit by the shell?

Calculate the time to reach sea level.

Vertical motion: $\quad s = ut + \frac{1}{2}gt^2$

$$80 = 0 \times t + \tfrac{1}{2} \times 9.81 \times t^2$$

$$t^2 = \frac{2 \times 80}{9.81} = 16.3 \qquad \therefore t = 4.0\,\text{s}$$

Considering the horizontal motion: $\quad s = v \times t = 400 \times 4.0 = 1600\,\text{m}$

You should now know:

- **objects on the Earth falling under gravity experience an acceleration of 9.81 m s^{-2}**
- **the motion of projectiles is motion in two dimensions — constant velocity in the horizontal direction and motion under gravity in the vertical direction**
- **when performing projectile calculations, each direction can be considered separately**
- **that the motions in the vertical and horizontal directions are linked by the time**

6 *Newton's laws of motion*

Newton's laws are required when we consider what happens to an object when forces act on the object.

Newton's first law

Every object continues in its state of uniform motion in a straight line unless it is compelled to change its state by external forces acting upon it.

The first law simply states what happens to an object when no force is applied: it remains stationary if it is stationary, or it continues to move in a straight line with no change in velocity, either in magnitude or direction.

See Topic 4, Section 1.2.

Newton's second law

The rate of change of momentum of a body is proportional to the total force acting on the body and takes place in the direction of the force.

This law tells us what happens to a body when a force is applied. This law is of more use when written in the more familiar form:

force = mass × acceleration

1 N is that force which causes a mass of 1 kg to accelerate at 1 m s^{-2}

When using this equation, if the mass is in kilograms and the acceleration in metres per second^2, then the units of force are newtons.

Newton's third law

If a body A exerts a force on body B, then body B must exert an equal and opposite force on body A.

The third law states that forces cannot act alone and that, whenever a force acts, there must be an equal and opposite balancing force somewhere in the system. Sometimes it is not easy to see where the second force is located.

Examples

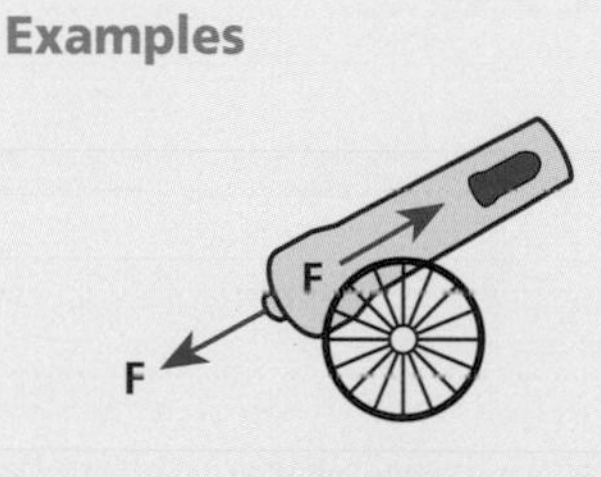

When a gun is fired, a force is exerted which accelerates the shell. An equal and opposite force is exerted on the gun, which accelerates the gun in the reverse direction.

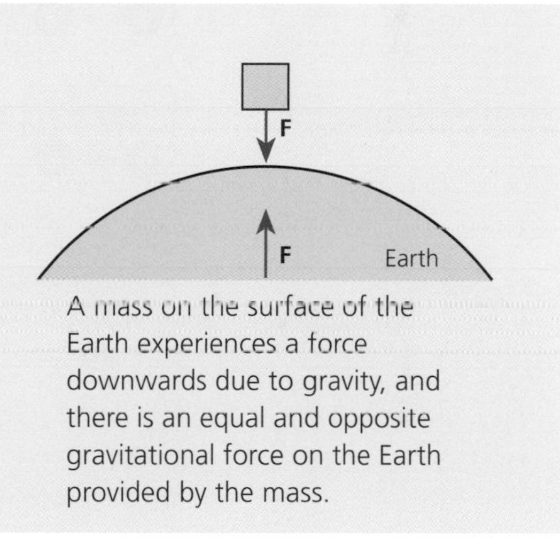

A mass on the surface of the Earth experiences a force downwards due to gravity, and there is an equal and opposite gravitational force on the Earth provided by the mass.

6.1 Free body view of forces

Because of the problems of locating all the forces in a complicated system, it is often useful in calculations to consider the system in terms of a set of free bodies.

Consider an engine pulling some wagons. The diagram below shows the forces on the couplings between each wagon.

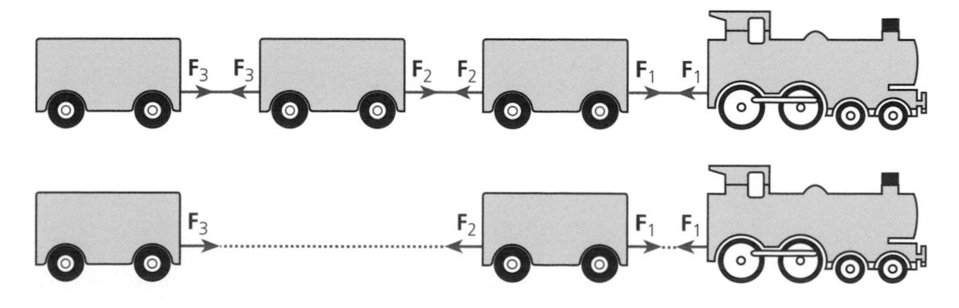

6.2 Mass and weight

The mass of a body is the property that resists motion. It has a constant value for a given body.

The weight of a body is the force that gravity exerts on it and is not constant. Weight changes as the gravitational force on the mass changes.

If an object has a mass of m kilograms then, using Newton's second law, the force on the mass due to gravity (the weight) is:

$$\text{force} = \text{weight} = m \times g \text{ N}$$

where g is the acceleration due to gravity.

Worked examples

Q1 A train of 20 trucks is accelerated at $1\,\text{m s}^{-2}$ by an engine. Each truck has a mass of 5 tonnes, i.e. 5000 kg. What is the force on the coupling between the engine and the first truck and on the coupling to the final truck?

All the trucks accelerate at the same rate, $1\,\text{m s}^{-2}$.

Using $F = m \times a$:

$$F_1 = 20 \times 5000 \times 1 = 100\,000\text{ N}$$

Last truck:

$$F_{20} = 5000 \times 1 = 5000\text{ N}$$

Q2 On the surface of the Earth, an astronaut in a space suit can jump to a height of 0.4 m. When on the surface of the Moon, he applies a force so that his initial velocity is the same as on the Earth. How high would he be able to jump on the surface of the Moon (g on the Moon = $1.6\,\text{m s}^{-2}$).

Using $v^2 = u^2 - 2gs$:

$$0 = u^2 - 2 \times 9.81 \times 0.4$$
$$u^2 = 2 \times 9.81 \times 0.4$$

On the Moon, $v^2 = u^2 - 2gs$:

$$0 = u^2 - 2 \times 1.6 \times s$$
$$u^2 = 2 \times 1.6 \times s$$

Hence: $2 \times 1.6 \times s = 2 \times 9.81 \times 0.4$

$$s = \frac{2 \times 9.81 \times 0.4}{2 \times 1.6} = 2.5\text{ m}$$

Q3 A shell of mass 2 kg leaves a gun at $54\,\text{m s}^{-1}$, having travelled along a horizontal barrel 4.8 m long. Calculate the magnitude of the acceleration and the force provided. What force is exerted on the gun?

Using $v^2 = u^2 + 2as$:

$$54^2 = 0^2 + 2 \times a \times 4.8$$

$$a = \frac{54^2}{2 \times 4.8} = 304 = 300\,\text{m s}^{-2}$$

$$\text{force} = m \times a = 2 \times 304 = 608 = 610\text{ N}$$

This is also the force on the gun but in the reverse direction.

You should now know:

- **the definitions and uses of Newton's laws of motion**
- **how to distinguish between mass and weight**

7 Forces in equilibrium

Application of Newton's first law means that if there are two or more forces acting on a body that is at rest, then the net force on the body must be zero. The forces are said to be in equilibrium.

In the following, we assume that all the forces lie in the same plane. This is all that is required for A-level.

7.1 Equilibrium of two forces acting at a point

When an object is in equilibrium and is acted on by two forces, the two forces must act at the same point, be equal in magnitude and opposite in direction.

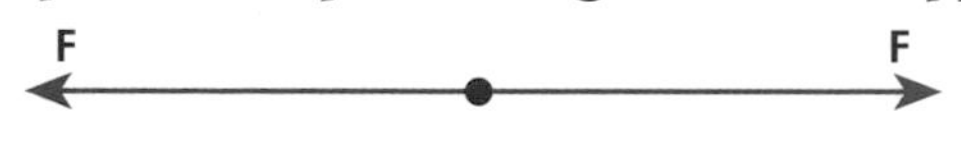

7.2 Equilibrium of three or more forces acting at a point

See Section 1.2 on the addition of vectors.

When an object is in equilibrium and is acted on by three or more forces at the same point, then the forces drawn as a vector diagram must form a closed polygon.

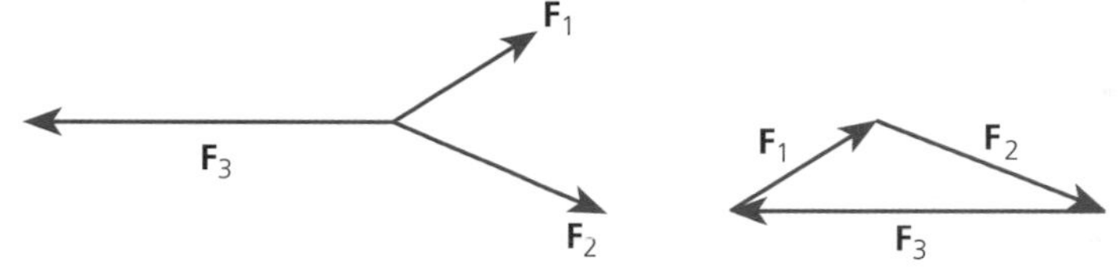

or

If the forces are resolved into horizontal and vertical components, the sum of the forces in the two directions must be zero.

7.3 Forces not acting at the same point

When the forces do not act at the same point, there is a possibility of rotation as well as linear motion.

The moment of a force

The moment of a force is defined as the force × the perpendicular distance from the line of action of the force to the axis of rotation.

Note: in diagram (b), when the force is at an angle to an arm, the perpendicular distance must be used to calculate the moment.

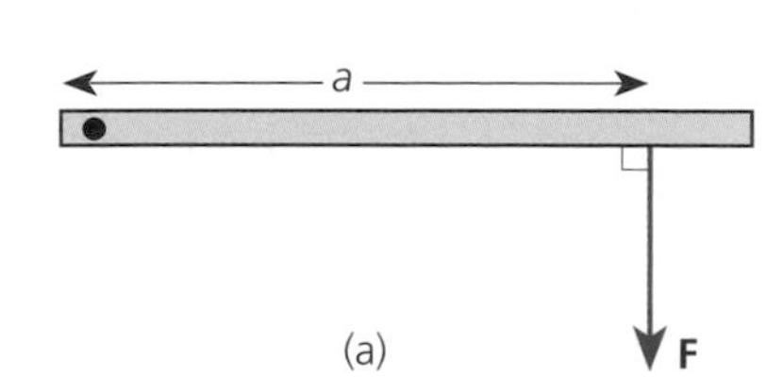

(a)

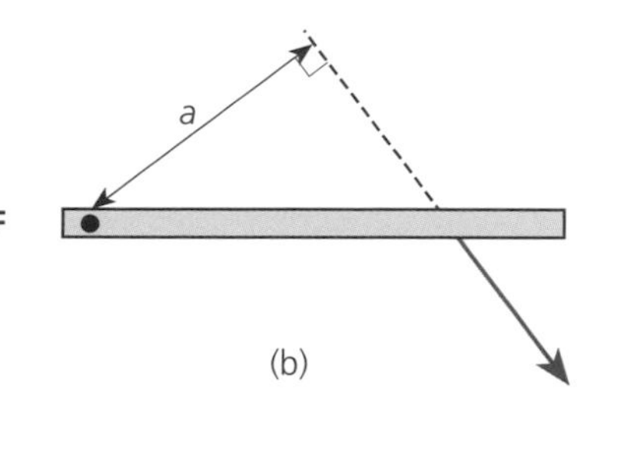

(b)

In both cases the moment $= F \times a$.

Couples

Rotating machines such as electric motors and car engines generate torque. The greater the torque generated, the more powerful is the rotating machine.

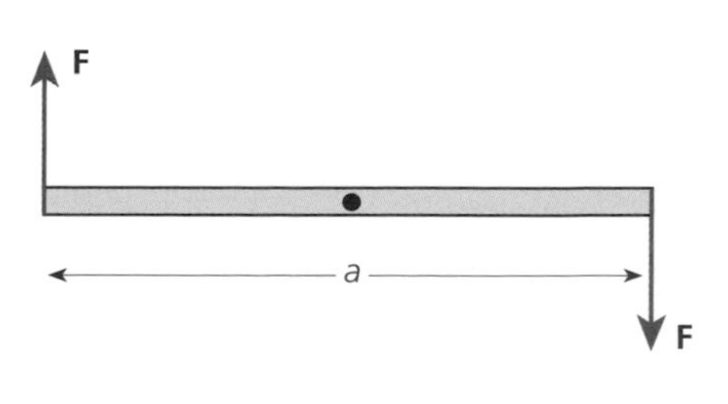

A couple is a special type of moment and consists of two parallel forces of the same magnitude acting in opposite directions. If the total moment of the system is taken about a central point, it has a total value $F \times a$, where a is the separation of the two forces. In this case, the magnitude of the total moment is called the torque.

$$\text{total moment} = F \times \frac{a}{2} + F \times \frac{a}{2} = F \times a = \text{torque}$$

7.4 Equilibrium of large objects

When forces are applied to large objects and the forces do not act at a single point, then there are two conditions for equilibrium:

1 **The vector sum of the forces must be zero, as above for a point object.**

2 **The clockwise moments about any point must equal the anticlockwise moments.**

Worked examples

Q1 A 3 kg mass is supported by two strings as shown. Calculate the tension forces in the two strings.

Convert the mass in kg to obtain the force downwards in N.

Resolving the forces vertically:

$$T_1 \times \cos 20 + T_2 \times \cos 10 = 3 \times 9.81 \quad (1)$$

Resolving horizontally:

$$T_1 \times \sin 20 - T_2 \times \sin 10 = 0 \quad (2)$$

From (2): $T_1 = T_2 \dfrac{\sin 10}{\sin 20} = T_2 \dfrac{0.174}{0.3442} = T_2 \times 0.51$

Substitute in (1): $T_2 \times 0.51 \times \cos 20 + T_2 \times \cos 10 = 3 \times 9.81$

$T_2 \times 0.48 + T_2 \times 0.98 = 29.4 \qquad \therefore T_2 = 20\,\text{N}$

$T_1 = 20.1 \times 0.51 = 10\,\text{N}$

Q2 A mobile consists of the following elements.

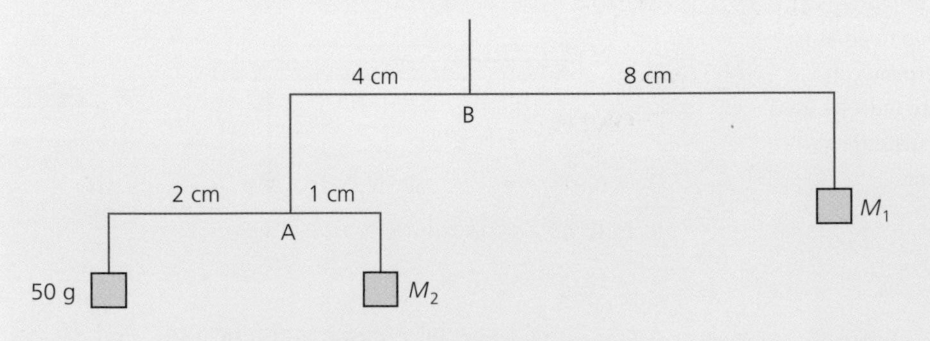

What must be the masses of M_1 and M_2 to keep the arms horizontal?

There is no need to convert to newtons here, as both sides of the equation would be multiplied by 9.81 m s^{-2} and would cancel.

Taking moments about A: $50 \times 2 = M_2 \times 1 \qquad M_2 = 100\,\text{g}$

Total force down at A: $50 + 100 = 150\,\text{g}$

Taking moments about B: $150 \times 4 = 8 \times M_1$

$$\therefore M_1 = \frac{150 \times 4}{8} = 75\,\text{g}$$

Convert cm to m to obtain the SI unit of torque.

Q3 Two equal and opposite forces act as shown below. Calculate the torque generated.

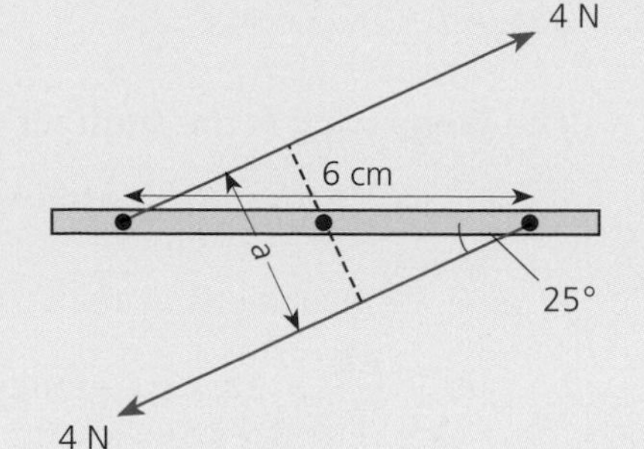

torque = force × perpendicular distance = $4 \times a$

But $\sin\theta = \frac{a}{6}$ $\therefore a = 6 \times \sin 25 = 2.53 = 0.0253\,\text{m}$

torque = $4 \times 0.0253 = 0.10\,\text{N m}$

You should now know:

- **the definition of a moment and a couple**
- **the condition for the equilibrium of a point object when coplanar forces are applied**
- **the conditions for the equilibrium of a large object when coplanar forces are applied**

8 Motion in a circle

Remember, velocity is a vector quantity and so has both magnitude and direction.

When an object moves along a curved path, the direction of motion of the object changes, hence the velocity must be changing.

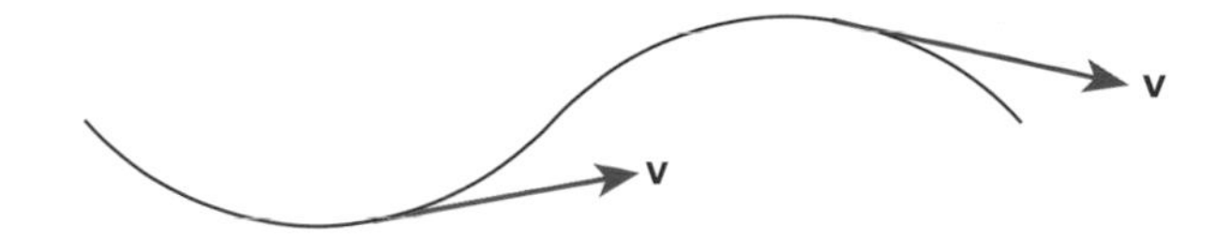

Consider the case of an object moving in a circle (fixed radius) with a constant speed. The velocity changes because the direction of motion of the velocity changes.

8.1 Angular velocity

In order to measure how fast an object moves around the circle, we introduce a new parameter called angular velocity, often given by the Greek symbol $\boldsymbol{\omega}$.

angular velocity = the angle turned through per unit time

See Topic 2, Section 9.1. At the centre of a circle there are 2π radians.

In circular motion, angles are measured in radians (rad) and angular velocity in rad s^{-1}.

8.2 Tangential velocity

The tangential velocity is related to the angular velocity by the equation:

tangential velocity = radius × angular velocity

8.3 The period and frequency

The period is the time taken for the object to move once around the circle or to travel through an angle of 2π radians.

If the angular velocity $= \boldsymbol{\omega} = \dfrac{\text{angle turned through}}{\text{time taken}}$

$$\omega = \frac{2\pi}{T}$$

where T is the period.

The frequency is the number of revolutions per second.

It takes T seconds to make one revolution.

Number of revolutions in 1 s = frequency = $\frac{1}{T}$

$$\therefore \omega = \frac{2\pi}{T} = 2\pi \times \text{frequency}$$

8.4 Centripetal acceleration in circular motion

Remember, Newton's first law states that if no force is applied the object just moves in a straight line.

Newton's first and second laws tell us that to change the state of motion of an object a force must be applied. Motion in a circle requires the state of motion of an object to be changed: the velocity must change, not in magnitude but in direction.

In the case of circular motion, the force must be applied towards the centre of the circle: a centripetal force.

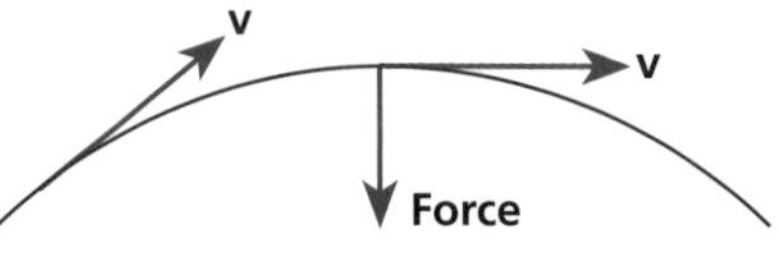

Newton's second law.

If the object has a mass, then the mass will have an acceleration in the direction of the force.

This is the **centripetal acceleration**, which equals $\omega^2 r = \frac{v^2}{r} = v\omega$

Using force = mass × acceleration:

$$\text{force} = m\omega^2 r = \frac{mv^2}{r} = mv\omega$$

8.5 Angular acceleration

If an object moves around the circle with increasing speed, **the change in angular velocity with time is the angular acceleration.**

Angular acceleration is measured in rad s^{-2}. There is also an acceleration parallel to the circumference of the circle.

tangential acceleration = radius × angular acceleration

Worked examples

Q1 Convert 45°, 90° and 110° into radians.

$360° = 2\pi$ rad

$$1° = \frac{2\pi}{360} \text{ rad}$$

$$45° = \frac{45 \times 2\pi}{360} = 0.79 \text{ rad}$$

$$90° = \frac{90 \times 2\pi}{360} = 1.6 \text{ rad}$$

$$110° = \frac{110 \times 2\pi}{360} = 1.9 \text{ rad}$$

Remember to divide the diameter by 2 to obtain the radius.

Q2 When travelling along the motorway, a car engine does 2800 revolutions per minute. What is its frequency of rotation and angular velocity? The flywheel at the back of the engine has a diameter of 35 cm. What is the centripetal acceleration of a point on the rim?

$$2800 \text{ rev min}^{-1} = \frac{2800}{60} \text{ rev s}^{-1} = 47 \text{ rev s}^{-1}$$

$$\omega = \frac{2\pi \times 2800}{60} = 290 \text{ rad s}^{-1}$$

$$\text{centripetal acceleration} = \omega^2 \times r = 290^2 \times \frac{0.35}{2} = 1.5 \times 10^4 \text{ m s}^{-2}$$

Q3 The Moon orbits the Earth once in 28.3 days. The Moon has a mass of 7.35×10^{22} kg and the radius of its orbit about the Earth is 3.84×10^8 m. Calculate the acceleration of the Moon towards the Earth and the gravitational force that the Earth must exert on the Moon to keep it in its orbit.

$$\text{period} = 28.3 \times 24 \times 60 \times 60 = 2.45 \times 10^6 \text{ s}$$

$$\omega = \frac{2 \times \pi}{T} = \frac{2 \times \pi}{2.45 \times 10^6} = 2.57 \times 10^{-6} \text{ rad s}^{-1}$$

$$\text{centripetal acceleration} = \omega^2 r = (2.57 \times 10^{-6})^2 \times 3.84 \times 10^8$$
$$= 2.54 \times 10^{-3} \text{ m s}^{-2}$$

$$\text{force} = m\omega^2 r = 7.35 \times 10^{22} \times 2.54 \times 10^{-3} = 1.86 \times 10^{20} \text{ N}$$

You should now know:

- **the definitions of the period, frequency, angular velocity and angular acceleration**
- **why there is a force and an acceleration towards the centre of a circle in circular motion**
- **expressions for the force and the centripetal acceleration in terms of the angular velocity and the radius of the circle**

TOPIC 4 Momentum and energy

1 Momentum

A body of mass m moving with a velocity $\mathbf{v}$ has momentum $\mathbf{p} = m \times \mathbf{v}$.

Momentum is a vector quantity, since velocity is a vector quantity.

1.1 Change in momentum

The momentum of a mass m can change when the velocity changes and, since velocity is a vector quantity, this change in velocity can be due to a change in the magnitude, direction or a mixture of both.

When a mass m with velocity $\mathbf{v}$ strikes a wall and bounces back with the same velocity, the change in momentum is:

total change of momentum = final momentum − initial momentum

$= -m\mathbf{v} - m\mathbf{v}$

$= -2m\mathbf{v}$

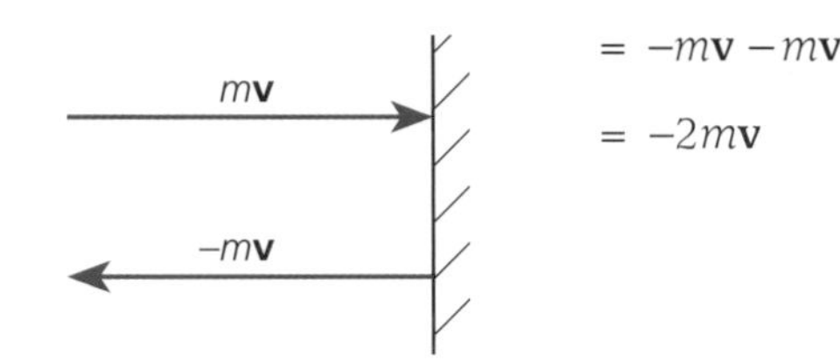

1.2 Newton's second law

Newton's second law states:

$$\mathbf{F} = m\mathbf{a}$$

So if the mass is constant:

$$\mathbf{F} = m\frac{\Delta \mathbf{v}}{\Delta t} = \frac{\Delta(m\mathbf{v})}{\Delta t} = \frac{\Delta \mathbf{p}}{\Delta t}$$

$\Delta \mathbf{p}$ is the change in momentum.

Force equals rate of change of momentum.

1.3 Impulse

Rearranging $\mathbf{F} = \frac{\Delta \mathbf{p}}{\Delta t}$ to $\mathbf{F}\Delta t = \Delta \mathbf{p}$ gives a new quantity, $\mathbf{F}\Delta t$, called the impulse of a force.

$$\textbf{impulse} = \mathbf{F}\Delta t = \Delta \mathbf{p}$$

Examples

A rocket engine: a rocket engine takes a certain mass of gas, which is initially at rest in the fuel tanks and, after heating the gas, ejects the same mass of gas out of the jet at a constant high velocity and at a constant rate. The force provided is obtained by calculating the change in momentum per second.

A jet engine: an aircraft jet engine does the same as a rocket engine except that air enters with an initial velocity and so has an initial momentum.

A tennis ball: a racquet striking a tennis ball provides a force for a very short period of time Δt. If it is assumed that the force is constant for this short time, the momentum change is given by the impulse.

1.4 Impulse and a changing force

When the force changes, the impulse can be obtained by drawing a force–time graph.

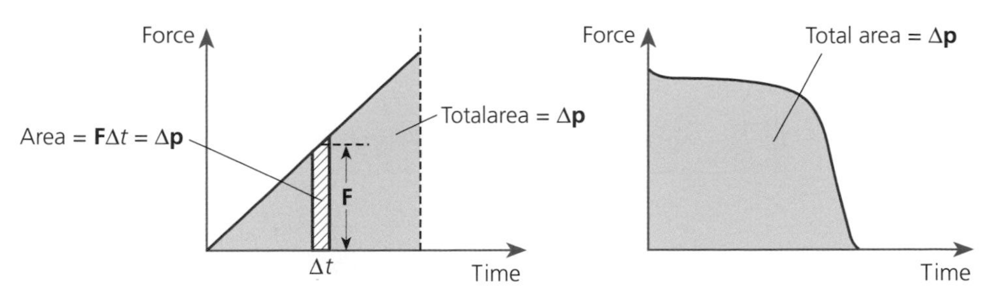

The total area under a force–time graph is the change in momentum.

In examples 1 and 3 we will take the momentum as positive when moving towards the ground or the wall.

The negative sign indicates that the change in momentum is in the opposite direction to the initial velocity.

Worked examples

Q1 A tennis ball of mass 55 g falls to the ground from a height of 1.0 m. What is its velocity on reaching the ground? It bounces back with its speed reduced by 10%. What is the change in momentum?

Calculate the velocity on striking the ground:

$v^2 = u^2 + 2gs \quad v^2 = 2 \times 9.81 \times 1.0 \quad v^2 = 19.6 \quad v = 4.4\ \text{m s}^{-1}$

velocity on leaving the ground $= 4.4 \times \dfrac{90}{100} = 4.0\ \text{m s}^{-1}$

change of momentum = final momentum − initial momentum

$= -0.055 \times 4.0 - 0.055 \times 4.4 = -0.46\ \text{kg m s}^{-1}$

Q2 A garden hose ejects water at a rate of 25 litre min^{-1} and a speed of 14 m s^{-1}. What force does the hose exert on the person holding it if the speed of water in the hose is 1 m s^{-1}?

Calculate the mass of water ejected per second:

1 litre = 1000 cm^3 = 1000 g = 1 kg

$\text{force} = \Delta(m \times v) = \dfrac{25 \times 14}{60} - \dfrac{25 \times 1}{60} = 5.4\ \text{N}$

Q3 A car collides with a wall, bouncing back with a small velocity. The time of the collision is 0.1 s. The mass of the car is 1050 kg, its velocity on striking the wall is 17 m s^{-1}, and it bounces back with a velocity of −0.5 m s^{-1}. Calculate the average force provided by the wall during the collision. What is its direction?

Calculate the momentum change at the wall:

$\Delta\mathbf{p}$ = final momentum − initial momentum

$= -1050 \times 0.5 - 1050 \times 17$

$\Delta\mathbf{p} = -18\,400\ \text{kg m s}^{-1}$

Use $\Delta\mathbf{F} \times \Delta t = \Delta\mathbf{p}$:

$\Delta\mathbf{F} = \dfrac{\Delta\mathbf{p}}{\Delta t} = -\dfrac{18\,400}{0.1} = -1.84 \times 10^5\ \text{N}$

The force is in the same direction as the final velocity.

You should now know:

- the definitions of momentum and impulse
- Newton's second law as $\mathbf{F} = \frac{\Delta p}{\Delta t}$.
- momentum as the area under a force–time graph
- how to explain the application of the above to practical situations

2 Work and energy

When an object is moved through a distance by a force, work is done.

The magnitude of the work done is given by force × distance moved, assuming the force is in the same direction as the distance moved.

Look back at resolving a vector into two at right angles (Topic 3, Section 1.3).

If the force and distance moved are not in the same direction, then the force must be resolved into a component along the direction in which the object moves.

Note that the force at right angles to the direction of motion, $F \sin\theta$, does no work.

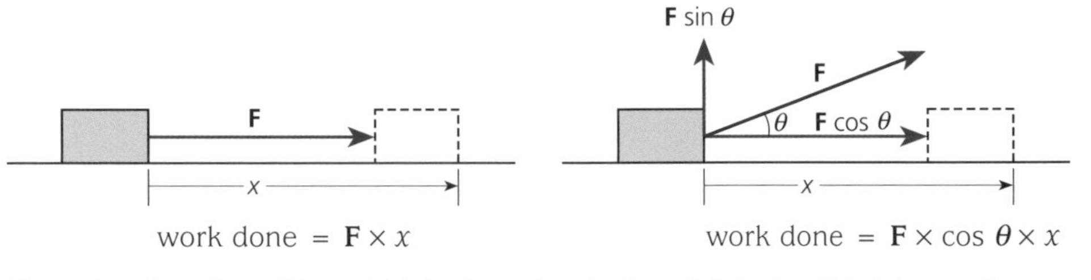

work done = $\mathbf{F} \times x$ work done = $\mathbf{F} \times \cos\theta \times x$

The units of work are Nm, which is given the single unit J, joules. Work is a scalar quantity.

2.1 Work done by a varying force

If the force varies, then a graph of force against distance enables the total work done to be calculated from the area under the graph.

Use the same method as for obtaining the momentum from a force–time graph.

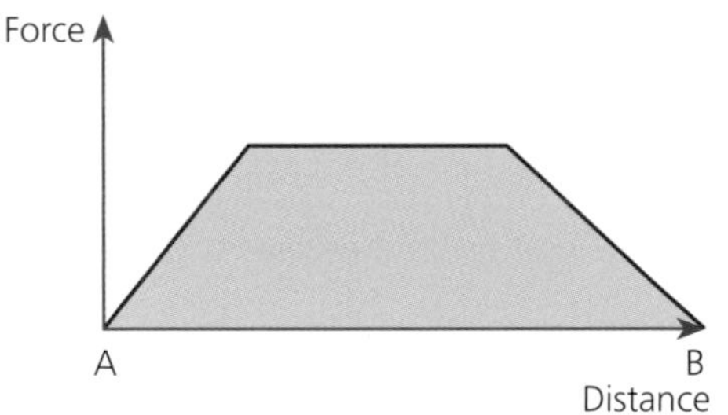

Work done in moving from A to B is the area under the graph.

When work is done *on* an object its energy increases.

When work is done *by* an object its energy decreases.

Energy has a number of different forms: mechanical energy, heat energy, chemical energy etc., all of which can do work.

2.2 Mechanical energy

Mechanical energy is the one most familiar to us. It comes in two forms: kinetic energy possessed by an object that is moving, and potential energy possessed by an object depending on its position. On many occasions, objects transfer energy between these two forms.

2.3 Kinetic energy

The kinetic energy, or KE, of a moving object depends on its mass and velocity, and is the work done to accelerate a mass m from zero to a velocity v.

kinetic energy = $\frac{1}{2}mv^2$

2.4 Potential energy

Potential energy, or PE, is the energy possessed by an object by virtue of its position in some force field, such as a gravitational field, an electric field or a magnetic field.

The potential energy at any point is measured with respect to some zero position, which can be chosen to suit the problem being considered. However, in many examples, we only need consider the potential energy change in moving from one point to another in the field.

2.5 Gravitational potential energy

When a mass m moves up and down above the surface of the Earth, there is a change in gravitational potential energy.

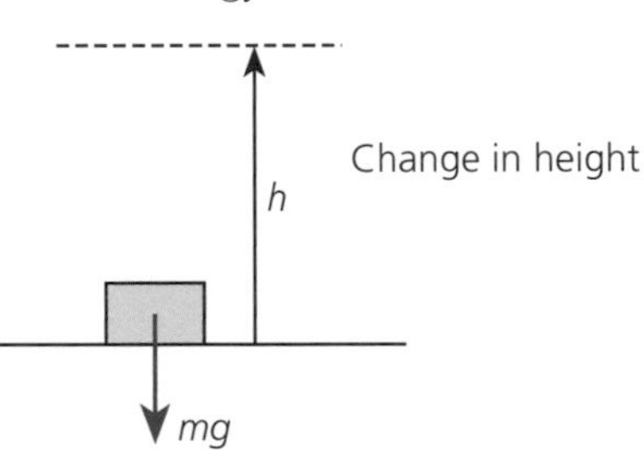

the change in gravitational potential energy = $m \times g \times$ the change in height

Moving up, the PE increases; moving down, the PE decreases.

How quickly energy is supplied, i.e. the power, is very important. For example, it dictates how quickly your kettle will boil water.

2.6 Power

$$\text{power} = \frac{\text{work done}}{\text{time taken}} = \frac{\text{the energy used}}{\text{time taken}}$$

The units of power are joules/second, or watts: $1\ \text{J s}^{-1} = 1\ \text{W}$.

When considering the conservation of energy in nuclear physics, we must consider the mass of an object in terms of energy using the equation $\Delta E = \Delta mc^2$, since mass and energy are equivalent (see Topic 9, Section 9).

2.7 Conservation of energy

Energy cannot be created or destroyed. It is just converted from one form to another.

When a mass falls to the ground, its initial PE is converted into KE as it falls and, when it strikes the ground, the KE is converted into heat energy, sound energy etc.

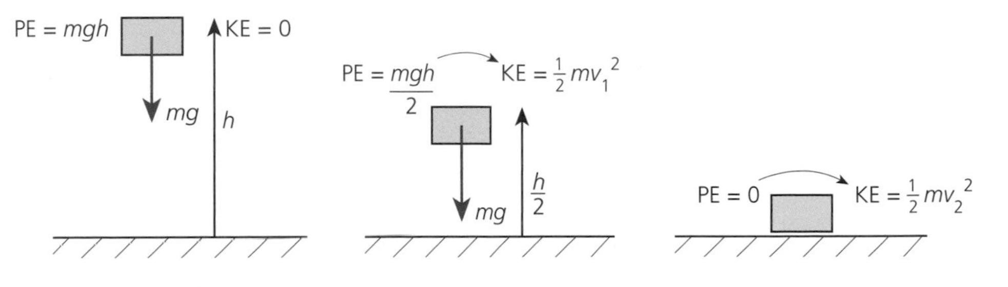

Worked examples

Q1 A woman has a mass of 50 kg and climbs up two flights of stairs through a total vertical distance of 7 m. How much work has she done? What is the change in her PE?

change in PE $= mg \times \Delta h$

$= 50 \times 9.81 \times 7.0 = 3400\ \text{J}$

Q2 A block is pulled by a rope applying the force shown. The block moves through a distance of 22 m.

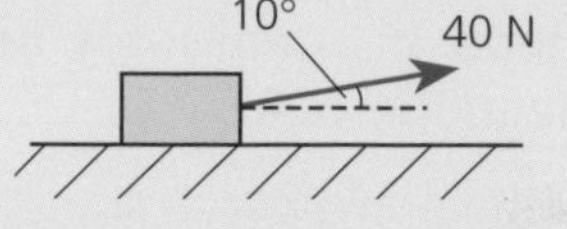

How much work has been done? What single force acting in the direction of motion would do the same amount of work?

force in the direction of motion = F $\cos\theta = 40 \times \cos 10$

Calculate the work done: $40 \times \cos 10 \times 22 = 867 = 870$ J

With just a horizontal force: $867 = F \times 22$

$F = 39$ N

Q3 A stone of mass 150 g is thrown vertically upwards with a velocity of 12 m s^{-1}. Calculate the initial kinetic energy and the maximum height reached. At what height will the velocity have dropped to 4 m s^{-1}?

Remember, to convert the mass to kg.

Calculate the initial KE $= \frac{1}{2}mv^2$:

$$\text{KE} = \frac{1}{2} \times 0.150 \times 12^2 = 10.8 \text{ J}$$

At maximum height, final PE = initial KE:

Note that we need not substitute m, as it appears on both sides of the equation.

$$\cancel{m}gh = \frac{1}{2}\cancel{m}v^2$$

$$h = \frac{v^2}{2g} = \frac{12^2}{2 \times 9.81} = 7.3 \text{ m}$$

$$\cancel{m}gh = \text{change in KE} = \frac{1}{2}\cancel{m}12^2 - \frac{1}{2}\cancel{m}4^2$$

$$h = \frac{1}{2 \times g}(144 - 16) = 6.5 \text{ m}$$

You should now know:

- **the definition of work and its units and how to calculate the work done**
- **the relationship between work, energy and power**
- **mechanical PE and KE and the law of conservation of energy**

3 *The collision of two particles in one dimension*

When two particles collide, the resulting motion can be calculated by the application of two conservation laws.

Remember, momentum is a vector quantity and, in order to apply the law, we must decide in which direction momentum is positive.

1 Conservation of momentum: the total momentum before the collision is equal to the total momentum after the collision.

This is a vector equation. This law always applies provided no external forces are acting.

2 Conservation of energy: the total energy must also be constant.

This is a scalar equation.

3.1 Collision types

If the initial energy of colliding particles is only kinetic, then after the collision there are a number of different results depending on what happens to the KE during the collision.

- **Elastic collisions** — the total KE of the particles prior to the collision is equal to the total KE of the particles after the collision. No collision is truly elastic, since in all collisions a fraction of the energy, no matter how small, is converted into some other form.

- **Inelastic collisions** — the total KE is not conserved and energy is released from the system in some other form such as heat, for example. There are two subdivisions of inelastic collisions:
  - **partially inelastic collisions**, in which *some* KE is converted into other energy forms
  - **totally inelastic collisions**, in which *all* the KE is converted into other forms
- **Super-elastic collisions** — the total KE is not conserved and energy is added to the system during the collision.

Other collisions can take place with $u_2 = 0$ or with u_2 moving to the left, in which case the momentum m_2u_2 would be negative.

3.2 An elastic collision

•→ m_1u_1	•→ m_2u_2	•→ m_1v_1	•→ m_2v_2	$u_2 < u_1$
Before the collision		***After the collision***		

Conservation of momentum

Let momentum to the right be positive:

$$m_1u_1 + m_2u_2 = m_1v_1 + m_2v_2$$

Conservation of energy

$$\tfrac{1}{2}m_1u_1^2 + \tfrac{1}{2}m_2u_2^2 = \tfrac{1}{2}m_1v_1^2 + \tfrac{1}{2}m_2v_2^2$$

Conservation of relative velocities

In a head-on collision, the relative velocity of approach of the two bodies is equal to the relative velocity of separation:

$$u_1 - u_2 = v_2 - v_1$$

3.3 An inelastic collision

In an inelastic collision, KE must be converted into some other form, since energy cannot be created or destroyed. It may not be easy to list all the changes in energy as a result of the collision or their relative magnitudes.

•→ m_1u_1	•→ m_2u_2	•→ m_1v_1	•→ + energy m_2v_2	$u_2 < u_1$
Before the collision		***After the collision***		

Conservation of momentum

Let momentum to the right be positive:

$$m_1u_1 + m_2u_2 = m_1v_1 + m_2v_2$$

Conservation of energy

$$\tfrac{1}{2}m_1u_1^2 + \tfrac{1}{2}m_2u_2^2 = \tfrac{1}{2}m_1v_1^2 + \tfrac{1}{2}m_2v_2^2 + \text{energy}$$

Example

A ball striking the ground and rebounding.

$\tfrac{1}{2}m_1u_1^2$ $\quad$ $\tfrac{1}{2}m_1v_1^2$ $\quad$ $v_1 < u_1$

energy (heat, sound...)

Remember, velocity is a vector quantity and requires an answer in both magnitude and direction.

Worked examples

Q1 A ball of mass 2 kg strikes a stationary ball of mass 3 kg with a velocity of 6 m s^{-1} and rebounds with a velocity of 0.4 m s^{-1}. What is the velocity of the second mass after the collision?

Conservation of momentum: $m_1u_1 + m_2u_2 = m_1v_1 + m_2v_2$

$$2 \times 6 + 0 = -2 \times 0.4 + 3 \times v_2$$

$$12 = -0.8 + 3v_2$$

$$3v_2 = 12.8$$

$v_2 = 4.3$ m s^{-1} in the same direction as u_1

Q2 A bullet of mass 14 g leaves a gun of mass 700 g with an energy of 630 J. Calculate the velocity of the bullet and the recoil velocity of the gun.

Find the velocity of the bullet: $\text{KE} = \frac{1}{2}mv^2 = 630 = \frac{1}{2} \times 0.014 \times v^2$

$$v^2 = 90\ 000$$

$$v = 300 \text{ m s}^{-1}$$

Using the conservation of momentum:

$$0 = m_1v_1 + m_2v_2$$

$$0 = 0.014 \times 300 + 0.700v_2$$

$$v_2 = \frac{-0.014 \times 300}{0.700} = -6 \text{ m s}^{-1}$$

The negative sign means the velocity of the gun is in the opposite direction to the bullet.

Q3 An object of mass 300 kg travelling at a speed of 10 m s^{-1} has a head-on collision with an object of mass 200 kg travelling at a speed of 10 m s^{-1}. What is the velocity of the two objects after the collision?

Using the conservation of momentum: $300 \times 10 - 200 \times 10 = 300v_1 + 200v_2$

The relative velocity is: $20 = v_2 - v_1$

$$1000 = 300v_1 + 200(v_1 + 20)$$

$$v_1 = -\frac{30}{5} = -6 \text{ m s}^{-1} \text{ and } v_2 = \frac{70}{5} = 14 \text{ m s}^{-1}$$

You should now know

- **the two conservation laws used in the collision of particles**
- **the definition of an elastic collision**
- **the difference between elastic and inelastic collisions**
- **the possible energy changes in inelastic collisions**
- **the difference between inelastic, partially inelastic and super-elastic collisions**

4 *The kinetic theory of gases*

In the following sections, we will be reviewing the physics of gases. There are two methods we can use to look at gases:

The Brownian motion experiment demonstrates that, in a gas, molecules are in random motion.

1 **Microscopic view**, in which we attempt to look at the behaviour of the gas in terms of the microscopic motions of the molecules — their velocities, energies etc. (the kinetic theory of gases).
2 **Macroscopic view**, in which we look at the behaviour of the gas in terms of macroscopic parameters such as pressure, temperature, volume etc., which we can measure in the laboratory.

Take care in this section to distinguish between atoms and molecules. For example, oxygen is a gas consisting of diatomic molecules (molecules that are two atoms joined together).

4.1 Definitions — macroscopic parameters

Mole: the amount of matter that contains as many atoms or molecules as there are in 12 g of isotope carbon-12. This number is the Avogadro constant $N_A = 6.022 \times 10^{23}$ mol^{-1}. The mole is a measure of the amount of a substance, specified in terms of a fixed number of atoms or molecules.

Atomic mass or **molar mass**: the mass of 1 mole of the substance, measured in units kg mol^{-1}.

Molar volume V_m: the volume occupied by 1 mole of a substance, measured in units m^3 mol^{-1}.

Relative atomic/molecular mass M_r: the mass of one atom/molecule relative to one-twelfth the mass of carbon-12. There are no units for this parameter.

Unified atomic mass unit (u): one-twelfth the mass of a single atom of carbon-12.

$$1\ \text{u} = \frac{0.012}{6.022 \times 10^{23} \times 12} = 1.66 \times 10^{-27}\ \text{kg}$$

Pressure: the perpendicular force a gas exerts per unit area, measured in units N m^{-2} or pascals, Pa.

Volume: measured in units m^3.

Temperature: measured in degrees kelvin/absolute units, K.

s.t.p.: standard atmospheric pressure and temperature — 1 atmosphere or 1.01×10^5 Pa, and 0°C or 273.15 K. Calculations on gases are often undertaken at s.t.p.

4.2 The kinetic theory of gases — the ideal gas

This considers a gas as molecules in random motion, and states that the pressure of a gas on the walls of a container is due to the collision of gas molecules with the walls. The force generated is associated with the rate of change of momentum of the gas molecules.

See Topic 3, Section 6 on Newton's second law of motion.

Assumptions of the kinetic theory for an ideal gas are:
- a gas consists of a large number of molecules
- the molecules are in rapid random motion
- collisions between molecules are elastic
- there are no attractive forces between molecules
- the molecules are small compared with the volume of the container
- when molecules collide, there are repulsive forces, but the time of the collision is negligible compared with the time between collisions

In a container, the average velocity of the molecules is zero. They travel in all directions.

The **mean speed** of the molecules, however, has a numerical value:

$$\bar{c} = \frac{c_1 + c_2 + c_3 + c_4 \ldots c_n}{N}$$

Take care here: mean-square speed means that you must square all the speeds first and then take the average. Note that the 2 is under the bar.

The **mean-square speed** also has a numerical value. This is the average of the speeds of all the molecules squared:

$$\overline{c^2} = \frac{c_1{}^2 + c_2{}^2 + c_3{}^2 + c_4{}^2 \ldots c_n{}^2}{N}$$

The kinetic theory also introduces a parameter called the **root-mean-square speed**, or the **r.m.s. value**, which is the square-root of the **mean-square speed**.

$$c_{\text{rms}} = \sqrt{\overline{c^2}}$$

4.3 Pressure of an ideal gas

You would not be expected to derive this at A-level.

The kinetic theory shows that the pressure of a gas on the walls of a container is given by the simple equation:

$$\text{pressure} = \tfrac{1}{3}\rho\overline{c^2}$$

where ρ is the density of the gas and $\overline{c^2}$ is the mean-square speed.

4.4 Internal energy of an ideal gas

In a real gas, there are attractive forces between molecules, and so the internal energy consists of kinetic energy plus potential energy due to molecular attractions.

The molecules of a gas possess kinetic energy as they move around. This is called the **internal energy** of the gas. As the gas is heated, the molecules gain energy and move around faster.

The **total internal energy** is given by the equation: $\boldsymbol{U = \frac{3}{2}nRT}$

where n is the number of moles, R is the ideal gas constant (8.31 J mol^{-1} kg^{-1}) and T is the absolute temperature in kelvin.

4.5 Kinetic energy of one molecule

The kinetic energy of 1 mole of an ideal gas is:

$$\text{KE} = \frac{3}{2}nRT \qquad n = 1$$

but in 1 mole there are N_A molecules, so the kinetic energy of 1 molecule is:

$$\frac{3}{2}\frac{R}{N_A}T = \frac{3}{2}kT$$

where $k = 1.38 \times 10^{-23}$ J K^{-1}. This is called the **Boltzmann constant** and is the gas constant for a single molecule.

Worked examples

Q1 The molecular mass of nitrogen, N_2, is 0.028 kg mol^{-1}. How many moles are there in 0.08 kg? What is the mass of each molecule and what is the relative molecular mass of N_2?

The number of moles $= \dfrac{0.08}{0.028} = 2.9$ moles

The mass of each molecule $= \dfrac{0.028}{6.02 \times 10^{23}} = 4.7 \times 10^{-26}$ kg

Relative molar mass = 28

Q2 Five molecules in a gas have speeds of 450, 360, 290, 380 and 420 m s^{-1}. Calculate the mean speed, the mean-square speed and the root-mean-square speed of the five molecules.

$$\bar{c} = \frac{450 + 360 + 290 + 380 + 420}{5} = 380\ \text{m s}^{-1}$$

$$\overline{c^2} = \frac{450^2 + 360^2 + 290^2 + 380^2 + 420^2}{5} = 1.47 \times 10^5\ \text{m}^2\,\text{s}^{-2}$$

$$c_{rms} = \sqrt{1.47 \times 10^5} = 384\ \text{m s}^{-1}$$

Q3 The density of hydrogen at s.t.p. is 9.0×10^{-2} kg m^{-3}. What is the root-mean-square speed of the hydrogen molecules?

$$\text{pressure} = \tfrac{1}{3}\rho\overline{c^2}$$

$$\overline{c^2} = \frac{3 \times p}{\rho} = \frac{3 \times 1.01 \times 10^5}{9.0 \times 10^{-2}} = 3.36 \times 10^6$$

$$c_{rms} = \sqrt{\overline{c^2}} = \sqrt{3.36 \times 10^6} = 1830\ \text{m s}^{-1}$$

You should now know:

- **the definitions of the mole, the Avogadro number, atomic/molar mass, relative atomic/molar mass, unified atomic mass**
- **the six basic assumptions of the kinetic theory for an ideal gas**
- **how to define mean-square speed and root-mean-square speed**
- **how to explain the origin of the internal energy for an ideal gas and a real gas**

5 Temperature

Temperature is measured by observing the variation in the physical property of a substance.

Temperature is a quantity used to describe how hot something is. It is a physiological response, since we talk about objects being hot or cold. It is a fundamental quantity, which cannot be described in terms of other fundamental quantities. Temperature describes how heat energy flows from one body to another. Heat energy always flows from a hot body to a cold body.

5.1 Zeroth law of thermodynamics

Bodies in thermal equilibrium, i.e. when no heat flows between them, are at the same temperature.

5.2 Temperature scales

The thermodynamic scale is based on the properties of an ideal heat engine, material not required at A-level.

The triple point is a unique temperature at which water, ice and water vapour are in equilibrium.

The thermodynamic temperature scale — Kelvin scale

This is a theoretical scale, which does not depend on the properties of a particular substance. It is defined as $\frac{1}{273.16}$ of the temperature of the triple point of water. The fraction was chosen so that one degree on the thermodynamic scale is the same as one degree on the centigrade scale.

The Kelvin scale has two fixed points: absolute zero, 0 K, and the triple point of water, 273.16 K.

$$T(X) = 273.16\,\frac{X}{X_{tr}}$$

where X is the value of the property used to measure the temperature at T and X_{tr} is the value at the triple point.

When performing calculations at A-level, it is acceptable to use the number 273 to convert from temperatures in kelvin to temperatures in degrees Celsius/centigrade.

The Celsius scale

This is the Kelvin scale adjusted to make it agree with the centigrade scale.

temperature in degrees Celsius = temperature in kelvin − 273.15.

Hence on this scale, absolute zero is −273.15° C, the freezing point of water 0°C and the triple point of water 0.01°C.

The centigrade scale

This is a practical temperature scale based on two fixed points: the freezing point of water, 0°C, and the boiling point of water, 100°C. To a first approximation, temperatures measured on the Celsius and the centigrade scales agree.

Ideal gas scale

An ideal gas obeys the equation:

pressure × volume = constant × temperature

Hence, if the pressure and volume of an ideal gas are measured at an unknown temperature, and at the triple point, then:

$$\frac{(PV)_{\text{unknown}}}{T_{\text{unknown}}} = \frac{(PV)_{\text{tr}}}{273.16}$$

If the volume is the same in both measurements:

$$T_{\text{unknown}} = 273.16 \times \frac{P_{\text{unknown}}}{P_{\text{tr}}}$$

Real gases behave like ideal gases at low pressures. So, in practice, the above temperature measurements can be made by taking measurements with real gases and observing the value of the calculated temperature as the pressure reduces to zero, called extrapolating to zero.

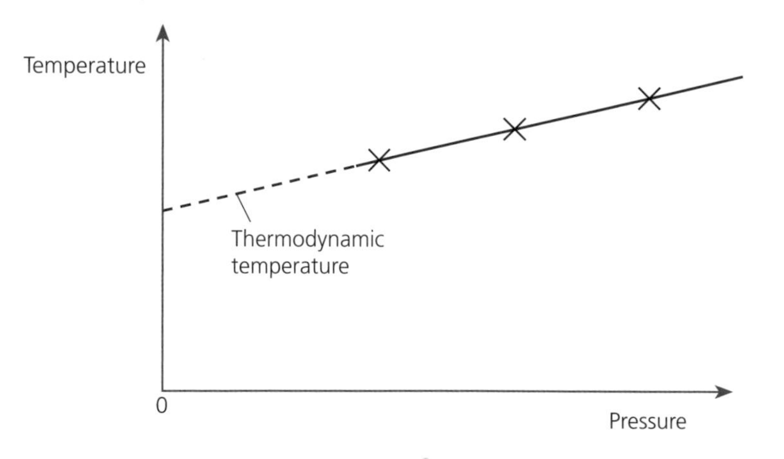

Practical scales of temperature

Because of the difficulty of making temperature measurements using an ideal gas, we make use of thermometers to measure temperature. A thermometer is a device that uses some property of a material that varies with temperature, such as changes in volume of a liquid, changes in the resistance of a metal or semiconductor, an e.m.f. generated between two metals, radiation from heated bodies etc.

Look in your textbook for the different designs of practical thermometers.

These devices then measure temperature between two fixed points, assuming the property changes either linearly between the two fixed points specified or according to a known equation.

A mercury-in-glass thermometer is assumed to be linear between the two fixed points, so an unknown temperature T is given by the equation:

$$T = \frac{L_T - L_0}{L_{100} - L_0} \times 100$$

where L_T, L_0 and L_{100} are the length of mercury at T, 0°C and 100°C.

The resistance of a thermistor varies non-linearly with temperature, but the resistance R_T, at any temperature T, is given by the equation:

$$R_T = a e^{\frac{b}{T}}$$

where a and b are constants.

Provided the variation with temperature of the property used in the thermometer is known, almost any property of a substance that varies with temperature can be used to construct a thermometer.

Worked examples

Q1 The resistance of a platinum wire at the triple point of water is 28 ohm and at the boiling point of water is 38 ohm. What is the temperature in kelvin of the boiling point of water using the platinum resistance scale?

$$T = 273.16 \times \frac{38}{28} = 370\text{ K}$$

Q2 A constant-volume gas thermometer uses oxygen as its working gas and measures the temperature of a hot liquid. At a pressure of 440 Pa, the temperature is calculated as 240.8 K, and at a pressure of 760 Pa, 241.4 K. What is the temperature of the liquid on the thermodynamic scale?

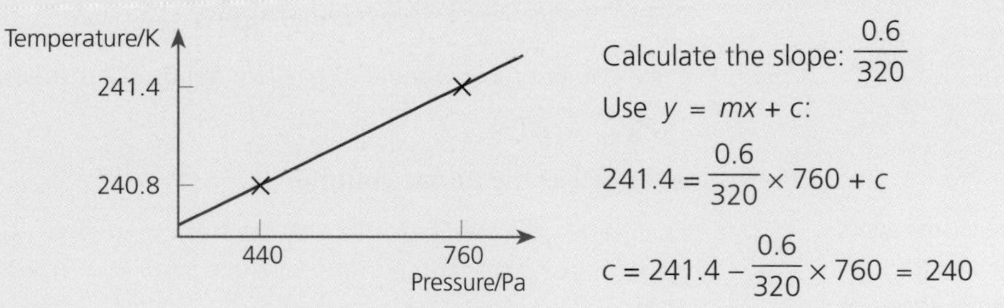

Calculate the slope: $\frac{0.6}{320}$

Use $y = mx + c$:

$$241.4 = \frac{0.6}{320} \times 760 + c$$

$$c = 241.4 - \frac{0.6}{320} \times 760 = 240$$

Q3 The cold junction of a thermocouple is at 0°C. With the hot junction at 600°C, the e.m.f. is found to be 60 mV. The hot junction is then placed in a hot liquid and the e.m.f. reads 47 mV. What is the temperature of the hot liquid measured by the thermocouple, assuming a linear scale?

$$T = \frac{E_T - E_0}{E_{600} - E_0} \times 600 = \frac{47 - 0}{60 - 0} \times 600$$

$T = 470$°C

Note that when the hot junction is the same temperature as the cold junction, the e.m.f. generated by the thermocouple is zero.

You should now know:

- **the difference between the Kelvin, Celsius and centigrade scales**
- **the triple point**
- **the ideal gas scale of temperature**
- **the ideal gas scale using real gases**
- **practical temperature scales and the use of thermometers**

6 *Gases — Boyle's law*

See Section 4 for the properties of an ideal gas.

When a pressure is applied to a gas, the volume decreases and, provided the gas approximates to an ideal gas, the volume V is inversely proportional to the pressure, P. This is Boyle's law.

$$P \propto \frac{1}{V} \qquad \text{or} \qquad PV = \text{constant}$$

It is also found that for an ideal gas:

We used this in Section 5 to obtain the ideal gas scale of temperature.

$V \propto T$ and $P \propto T$, which combined together give:

$$PV \propto T \qquad \text{or} \qquad \frac{PV}{T} = \text{constant}$$

From the above, we obtain the following equation, in which the subscripts 1 and 2 refer to the initial and final states of the ideal gas:

$$\frac{P_1V_1}{T_1} = \frac{P_2V_2}{T_2}$$

From the above equation to measure temperature on the Kelvin scale, T must be in kelvin.

6.1 The ideal gas equation

The more molecules there are in a gas, the greater the pressure, because there are more molecules colliding with the walls of the container per second. As a result, the rate of change of momentum will also be greater.

$$\frac{PV}{T} \propto n$$

where n is the number of moles. We can make this into an equation by introducing a suitable constant, R, the **molar gas constant**, which is 8.31 J K^{-1} mol^{-1}.

$$PV = nRT$$

If we consider just 1 mole of gas, we obtain the **universal gas law** equation:

$$PV_m = RT$$

where V_m is the **molar volume**.

In the examination, make sure the examiner knows what you mean by n and N.

The ideal gas law can also be written in terms of the number of molecules by remembering that the number of molecules $N = n \times N_A$.

$$PV = \frac{N}{N_a}RT \text{ but } \frac{R}{N_a} = k \text{ so } PV = NkT \qquad k = \text{Boltzmann's constant}$$

6.2 Real gases

Real gases differ from ideal gases in that there is an attractive force between the molecules, and the volume of the molecules is not small compared with the volume of the container. This becomes more important as the pressure of the gas increases and the volume decreases.

6.3 Isothermal and adiabatic changes

When the pressure and volume of a gas change, there are a number of occasions in physics when particular conditions apply.

Isothermal process

In an isothermal process, the temperature of the gas remains constant, so that T is the same at the end as it is at the beginning. Thermal energy must be

allowed to enter or leave the system, and the ideal gas equation then reduces to:

$$P_1V_1 = P_2V_2$$

Adiabatic process

In an adiabatic process the gas is in a closed container and no thermal energy is allowed to enter or leave the system. The above equation now becomes:

You will not need to know more than this equation for adiabatic processes.

$$P_1V_1^{\gamma} = P_2V_2^{\gamma}$$

where γ is a constant.

6.4 Work done when a gas is compressed

In order to compress a gas by a piston in a cylinder, an external force must be applied.

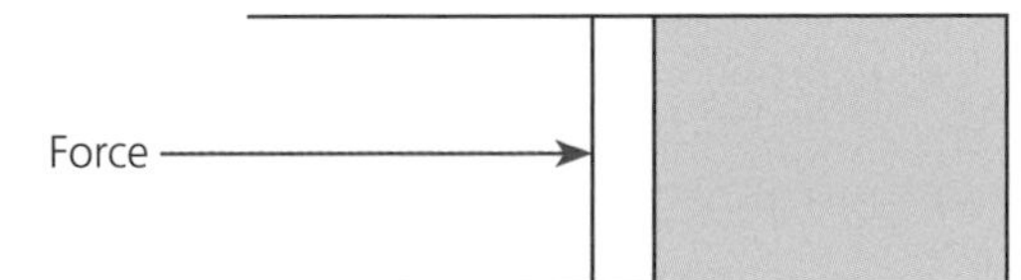

As the gas compresses, the force moves through a distance and work is done *on* the gas.

If the distance moved is small, the pressure can be assumed to be constant (not really the case).

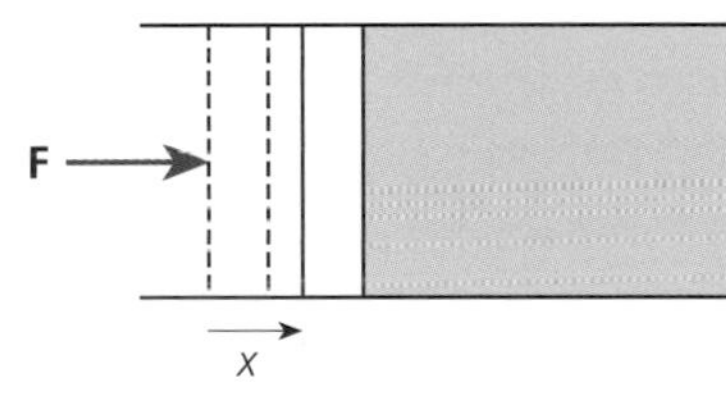

work done *on* the gas $= \mathbf{F} \times x$

6.5 Work done when a gas expands

When a gas expands work is done *by* the gas on the piston.

The same equations apply.

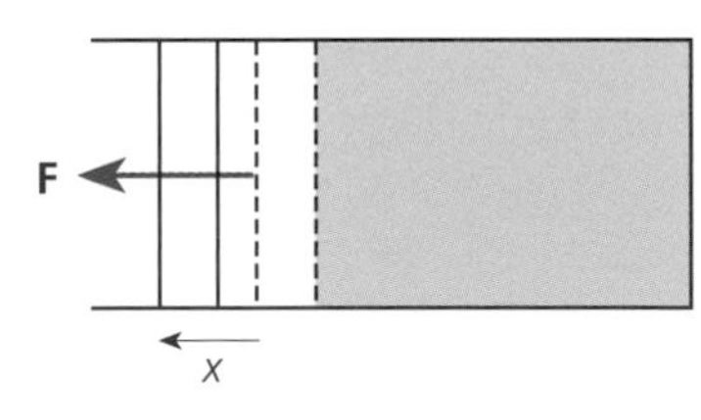

work done *by* the gas $= \mathbf{F} \times x$

1 atmosphere = 1.01×10^5 Pa

Worked examples

Q1 A car tyre is filled with air to a pressure of 1.5 atmosphere on a cold day when the temperature is 5°C. Assuming the volume of the air in the tyre remains constant, what would be the pressure, in atmospheres, in the tyre after travelling down the motorway, when the tyre temperature reached 35°C?

$$\frac{P_1V_1}{T_1} = \frac{P_2V_2}{T_2} \quad \text{but } V_1 = V_2 \quad \therefore \frac{P_1}{T_1} = \frac{P_2}{T_2}$$

$$\frac{1.5 \times 1.01 \times 10^5}{278} = \frac{x \times 1.01 \times 10^5}{308} \quad x = 1.7 \text{ atm}$$

Remember to convert temperature to kelvin.

Q2 What is the volume of 1 mole of an ideal gas at s.t.p.?

Using $PV = nRT$ $\quad V = \frac{nRT}{P} = \frac{1 \times 8.31 \times 273}{1.01 \times 10^5} = 2.25 \times 10^{-2}\ m^3$

$= 22.5 \times 10^{-3}\ m^3$

$= 22.5$ litres

The initial pressure in the pump is 1 atm.

Q3 A cyclist pumps up her cycle tyre with a pump, compressing the gas to $\frac{1}{4}$ of its volume. What would the final pressure of the gas be if the compression was undertaken (a) slowly, isothermally, (b) quickly, adiabatically? γ for air = 1.4.

Isothermally $\quad P_1V_1 = P_2V_2 \quad 1 \times V_1 = \frac{P_2V_1}{4} \quad P_2 = 4$ atm

Adiabatically $\quad 1 \times (V_1)^\gamma = P_2\left(\frac{V_1}{4}\right)^\gamma \quad P_2 = (4)^{1.4} \times 1$

$= 7.0$ atm

You should now know:

- **Boyle's law, the ideal gas equation and the universal gas law equation**
- **how to explain the difference between isothermal and adiabatic changes**
- **how to perform calculations using the two forms of the equation**
- **how to calculate the work when a gas changes in volume**

7 *Thermodynamics*

Thermodynamics is the physics of heat in motion.

7.1 Internal energy

In an ideal gas, the internal energy is simply the kinetic energy of the molecules as they move around, and is given by $U = \frac{3}{2}nRT$.

In a real gas or any other substance, the internal energy is the kinetic energy plus the potential energy of the atoms or molecules of the substance.

When the temperature changes, there is a change in internal energy when both the kinetic and the potential energy can change. When there is a change of state, say, solid to liquid, this takes place at a constant temperature, and only the potential energy of the system changes.

When a solid melts, **latent heat of fusion** is used. When a liquid vaporises, **latent heat of vaporisation** is used (see Topic 5, Section 3).

With a temperature change, $\Delta U = mc\Delta T$, where m is the mass, ΔT is the change in temperature and c is the specific heat capacity in units of J kg^{-1} K^{-1}.

With a change of state, $\Delta U = mL$, where m is the mass and L is the specific latent heat, in units of J kg^{-1}.

Internal energy can increase when energy is added to the system, or decrease when energy is given out by the system.

The first law of thermodynamics is simply the conservation of energy law applied to heat flow.

7.2 First law of thermodynamics

When the internal energy of a system **increases**, the energy can come from two possible sources:

1 Heat can **flow** into the system, and/or
2 Work can be done **on** the system by some external force.

Note the words in bold — they are important. A common error at A-level is just to talk about the changes in the energies without specifying the direction of the changes.

The first law of thermodynamics states that the *increase* in internal energy = heat flow into the system + work done *on* the system.

$\Delta U = Q + W$

Positive signs are associated with the processes defined above. Negative signs must be used for the reverse processes. Example: heat flow out from the system = $-Q$.

Isothermal and adiabatic changes

When the energy of the system changes, two important processes are possible.

1 **$Q = 0$. No heat flows in or out of the system, i.e. this is an adiabatic process. In this case the work done *on/by* the system equals the *increase/decrease* in the internal energy.**

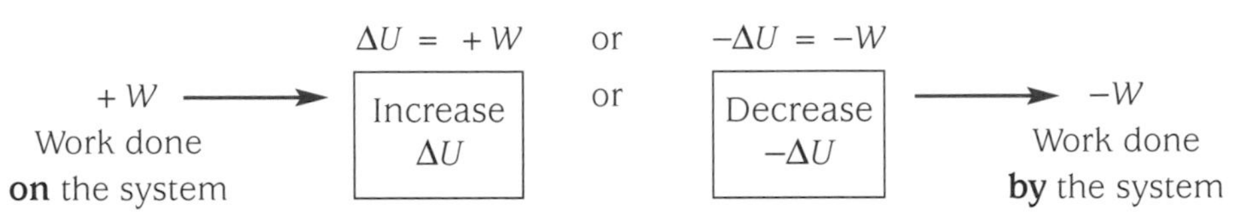

2 **$\Delta U = 0$. The internal energy is constant, i.e. this is an isothermal process. In this case, the heat *flow in/flow out* of the system equals the work done *by/on* the system.**

When values are inserted into the equation, heat *out* and work *done* by the system are negative.

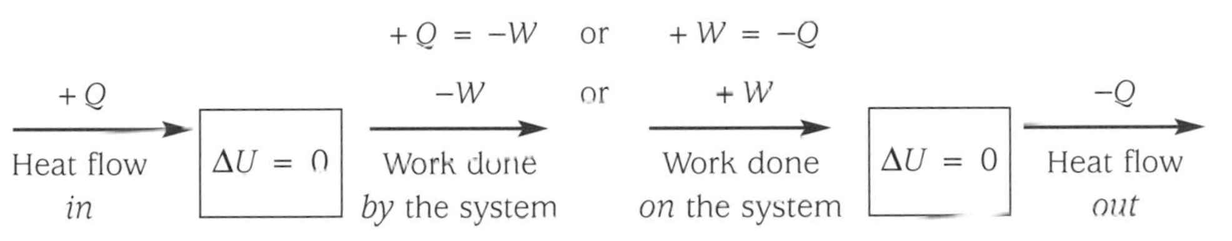

7.3 Second law of thermodynamics

This law deals with the conversion of heat energy into mechanical work and provides a theoretical limit to the amount of mechanical work we can obtain from a given amount of heat.

To convert heat into mechanical work, a **heat engine** is required. This is a device that takes a working substance around a cycle and draws heat from a hot reservoir, T_1, converts some of it into mechanical work, and ejects the remainder to a cold reservoir, T_2.

A car engine is one form of heat engine.

If the substance is taken around a cycle then the internal energy will not have changed, as the substance will end up in the same state as it starts, $\Delta U = 0$.

So for one cycle, $+Q = -W$. That is, the heat *flow into* the system = the work done *by* the system.

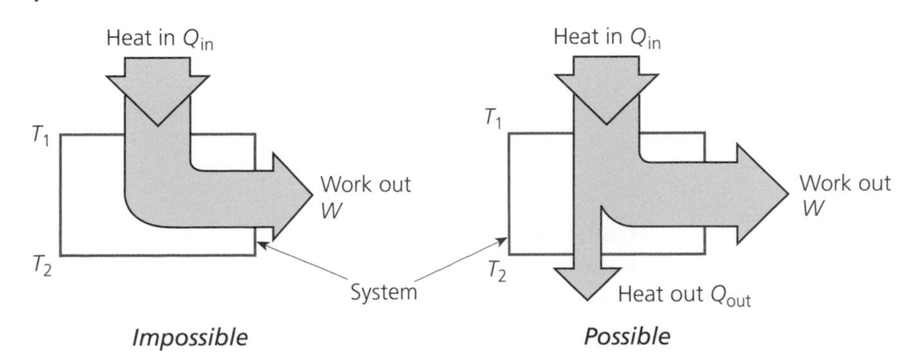

Maximum efficiency is 1 when $Q_{out} = 0$. This is impossible. Efficiencies are usually quoted in per cent, so multiply by 100.

... where T is the temperature in kelvin.

The second law of thermodynamics simply states that it is impossible to design a system in which $Q_{out} = 0$; we always get $\Delta Q = Q_{in} - Q_{out} = W$.

The efficiency of the conversion process

$$\text{efficiency} = \frac{\text{work out}}{\text{heat in}} = \frac{W}{Q_{in}} = \frac{Q_{in} - Q_{out}}{Q_{in}} = 1 - \frac{Q_{out}}{Q_{in}}$$

If a perfect heat engine, a Carnot engine, is designed, you can show that its efficiency is given by:

$$\text{efficiency} = \frac{Q_{in} - Q_{out}}{Q_{in}} = \frac{T_1 - T_2}{T_1} = 1 - \frac{T_2}{T_1}$$

All practical heat engines, such as car engines, gas turbines and steam engines, have much lower efficiencies than the ideal Carnot heat engine.

Worked examples

Q1 A kettle is filled with 1.5 kg of water at a temperature of 10°C. How much energy is required to convert all this water into steam at 100°C? Assume that no heat is lost.

$$\Delta U = mc\Delta T + mL = 1.5 \times 4200 \times 90 + 1.5 \times 2.26 \times 10^6$$

$$\Delta U = 5.67 \times 10^5 + 3.39 \times 10^6$$

$$\Delta U = 4.0 \times 10^6\ \text{J}$$

Q2 A ball of mass 0.1 kg at a height of 1.5 m falls to the ground and does not bounce. What are the changes in U, Q and W, and what is the increase in temperature of the ball? The ball is made of material with a specific heat of 6000 J kg^{-1} K^{-1}.

$$\Delta U = Q + W$$

But $Q = 0$ (no heat enters or leaves the system)

W = work done on the ball = $mgh = 0.1 \times 9.81 \times 1.5$

$$\therefore \Delta U = \cancel{m}c\Delta T = W = \cancel{m}gh$$

$$\Delta T = \frac{g \times h}{c} = \frac{9.81 \times 1.5}{6000} = 2.5 \times 10^{-3}\,°\text{C}$$

Q3 A heat engine operates between two temperatures, 350°C and 20°C. What would be the efficiency of a perfect engine between these two temperatures? If the real engine has an efficiency of 15%, how much heat is ejected to the cold reservoir when it does 4×10^2 J of work?

$$\text{efficiency} = 1 - \frac{T_2}{T_1} = 1 - \frac{293}{623} = 1 - 0.47 = 0.53 = 53\%$$

$$0.15 = \frac{W}{Q_{in}} \qquad \therefore Q_{in} = \frac{W}{0.15} = \frac{4 \times 10^2}{0.15} = 2.7 \times 10^3\ \text{J}$$

$$Q_{out} = Q_{in} - W = 2.7 \times 10^3 - 4 \times 10^2 = 2.3 \times 10^3\ \text{J}$$

You should now know:

- **how to define the internal energy of a system and calculate internal energy changes**
- **the first law of thermodynamics and its application**

- **the concept of a heat engine and the definition of its efficiency**
- **the second law of thermodynamics and its application to perfect and real heat engines**

8 Transfer of thermal energy

Wherever there is a temperature difference, heat will flow from the high temperature to the low temperature. There are three mechanisms by which heat flows:

- **conduction**
- **convection**
- **radiation**

Depending on the physical system concerned, one or all of these mechanisms may take place.

Remind yourself of the kinetic theory of gases — Section 4. Free electrons in a metal behave in a similar way to gas molecules when heated.

8.1 Thermal conduction

This is the mechanism of heat flow in solids, and it depends on the nature of the solid.

Metals consist of a lattice of positive ions through which free electrons can move. Heat energy is conducted through metals by two methods:

1. When heated, free electrons gain kinetic energy and drift through the ion lattice to the colder end.
2. Heat energy is transferred by the lattice ions, which vibrate with greater amplitude, passing these vibrations to neighbouring atoms.

In **insulators**, heat energy is transferred only by the lattice vibrations.

Remember, rate is per unit time.

Thermal conductivity

The rate of heat flowing down a metal bar can be calculated using the following equation:

$$\frac{Q}{t} = \lambda A \frac{(T_H - T_C)}{x}$$

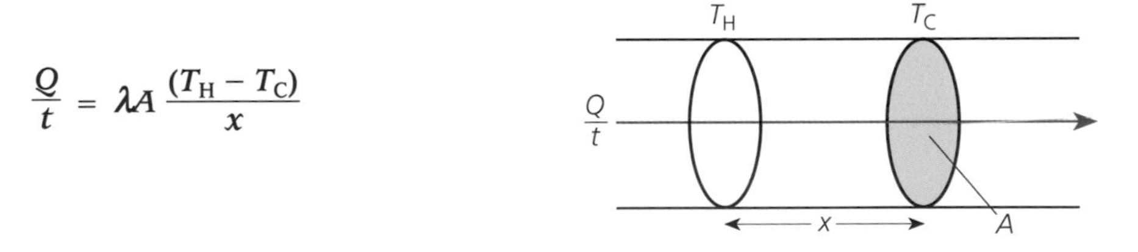

Q/t is the rate of heat flow. λ is a constant for the material of the bar, called the thermal conductivity, measured in units of W m^{-1} K^{-1}. A is the cross-sectional area in units of m^2. $\frac{(T_H - T_C)}{x}$ is the temperature gradient in K m^{-1} or °C m^{-1}.

The equation only works correctly for a lagged bar, where the temperature gradient is constant along the bar or when the system has a large area and a small thickness so that little heat is lost from the sides.

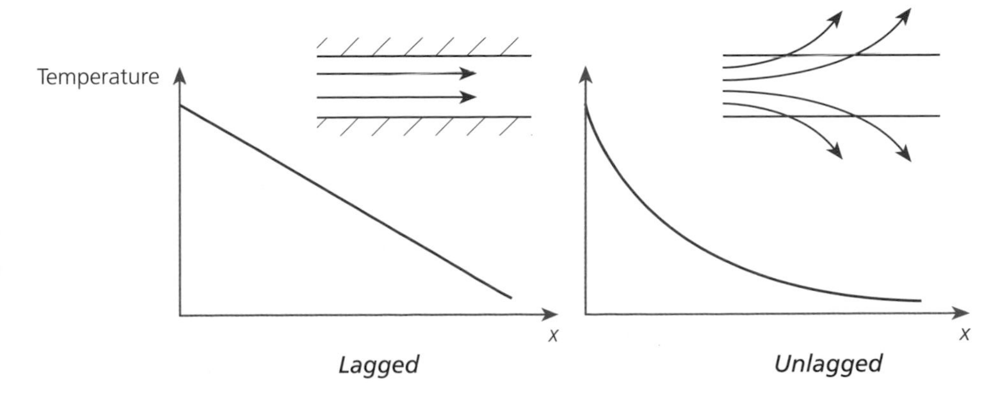

U-values

These take into account the thickness of the material by including it with λ. $U = \frac{\lambda}{x}$.

$$\frac{Q}{t} = \frac{\lambda}{x}A(T_H - T_C) = UA(T_H - T_C)$$

U-values measure the rate of heat flow through an area A of material for a certain temperature difference $(T_H - T_C)$. U-values have units of $W\,m^{-2}\,K^{-1}$.

8.2 Convection

This is a heat transfer process that takes place in fluids, liquids and gases owing to the fact that, in general, when a fluid is heated, its density decreases and the hot fluid rises. Hence, cold fluid sinks.

Mathematical equations calculating the rate of heat flow are difficult and are not part of an A-level physics course.

8.3 Radiation

The third method of heat transfer is radiation, heat transfer by electromagnetic waves.

Radiation will travel through a vacuum, whereas conduction and convection require a solid or a fluid for the transfer of heat.

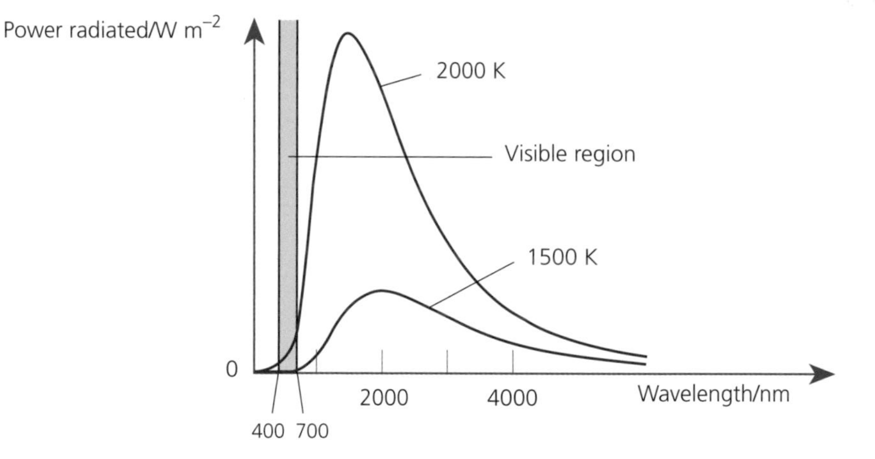

Infrared radiation is sometimes given the name heat radiation for obvious reasons.

The graph shows the power radiated as a function of wavelength from a heated body at two temperatures. The majority of the radiation is in the invisible, infrared region of the electromagnetic spectrum.

As the temperature of the heated body increases, we can make two observations:

1. **The intensity increases non-linearly, actually proportional to T^4; the intensity is proportional to the area under the curve.**
2. **The spectrum shifts slightly to shorter wavelengths, the short-wavelength tail moving into the visible region, which is why objects glow red when they are heated.**

Worked examples

Q1 A brick wall has a thermal conductivity of $0.6\,W\,m^{-1}\,K^{-1}$. In winter, the inside room temperature is maintained at 21°C and the outside is at 3°C.
What is the rate of heat flow through a wall 3 m by 6 m and 8 cm thick?

$$\frac{Q}{t} = \lambda A\frac{(T_H - T_C)}{x} = 0.6 \times 3 \times 6 \times \frac{(21-3)}{0.08} = 2400\ J\ s^{-1}$$

Q2 Suggest why modern stainless steel pans have metal handles, whereas older aluminium pans were often fitted with wooden handles.

Answer

The thermal conductivity of stainless steel is $50\,\mathrm{W\,m^{-1}\,K^{-1}}$, and that for aluminium is $205\,\mathrm{W\,m^{-1}\,K^{-1}}$. The rate of heat flow down aluminium is about four times that for stainless steel.

Q3 Why is efficient thermal insulation a good design feature for houses in both cold and hot countries? Why do houses in hot countries usually have small windows?

Answer

In both cases, there is a need to prevent heat flow. In cold climates, there is a need to stop heat flowing out. In hot climates, there is a need to stop heat flowing in. In hot countries, small windows mean less radiant heat energy from sunlight enters the house.

You should now know:

- **how heat is transferred by conduction, convection and radiation**
- **the conduction process in a metal and non-metal**
- **the conduction equation and the definition of thermal conductivity and *U*-value**
- **why convection takes place and heated objects glow red**

TOPIC 5 Solids and fluids

1 Deformation of solids

When a force is applied to a solid, it changes in shape, i.e. it deforms. The size and nature of the deformation produced depends on the material from which the solid is made. Solids are classified into broad groups, depending on the nature of the deformation.

1.1 Stress and strain

In order to measure the deformation of a solid when a force is applied, two parameters are defined:

1 **Stress** — the force applied to the solid divided by the area of cross-section over which the force is applied. $\frac{F}{A}$, units N m^{-2}.
2 **Strain** — the increase in length divided by the original length of the material, $\frac{\Delta L}{L}$, no units.

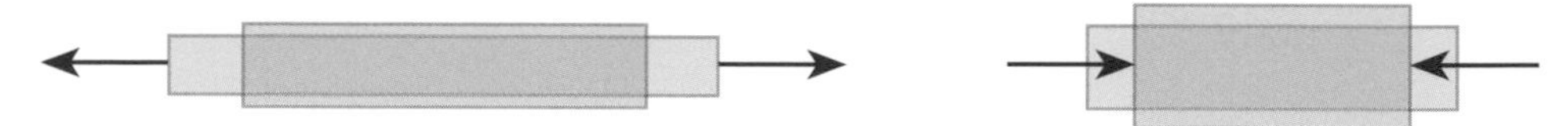

Tensile forces make the solid longer; compressive forces make the solid shorter.

Stress–strain graphs

The behaviour of materials subject to tensile or compressive forces is best observed with a stress–strain graph.

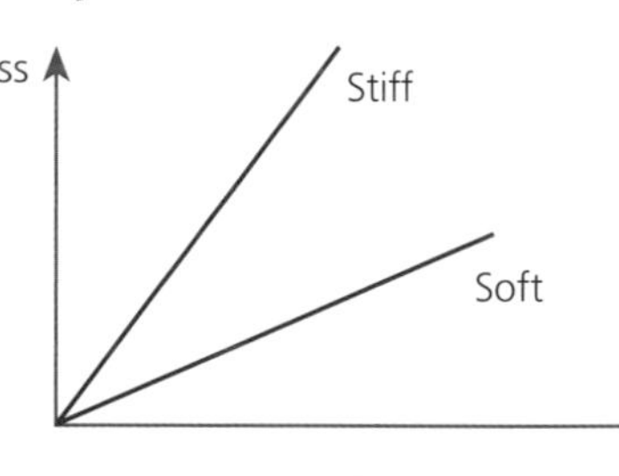

With a small force, the graph is linear and the material is said to obey **Hooke's law**.

The slope of the graph is a measure of the response of the material to the application of a force. A stiff material will have a steep slope.

Young's modulus, $E = \frac{\text{stress}}{\text{strain}}$ = the slope of the stress–strain graph

If the force is increased above a certain value, the graph becomes non-linear and its shape classifies solid materials into different types.

Limit of proportionality

The point at which the stress–strain graph becomes non-linear is called the limit of proportionality.

Take care here, as the limit of proportionality and the elastic limit may or may not be at the same point — it depends on the material. Usually, the elastic limit occurs at a higher stress than the limit of proportionality.

Elastic materials

An elastic material returns to its original shape when the stress applied is removed. The point on the stress–strain graph at which this ceases to be the case is called the elastic limit.

Plastic behaviour

After the elastic limit, materials may exhibit plastic behaviour in that, when the stress is removed, the material does not return to its original shape.

In the plastic region, a small increase in stress can cause a large increase in strain, and such materials are said to be ductile.

Breaking point

If the stress is further increased the material will break — the breaking point.

1.2 Behaviour of different materials

Materials may be classified into three main groups.

1 Crystalline materials in which the atoms have long-range order, forming large single crystal grains. When elastically deformed, the atoms move apart a small distance. Under plastic deformation, crystal planes slide over one another owing to the movement of dislocations.

2 Polymeric materials consist of long-chain molecules tangled together. When a stress is applied, the molecules stretch out in one direction. In general, a small stress produces a large extension. Cross-linking of these chains makes the material stiffer.

3 Amorphous materials have little or no crystalline structure; any structure that does exist extends over a very small distance. Amorphous solids such as glass show no plastic region and are said to be **brittle**.

Some polymeric materials show small regions of crystal structure.

PL = plastic limit
EL = elastic limit
BP = breaking point

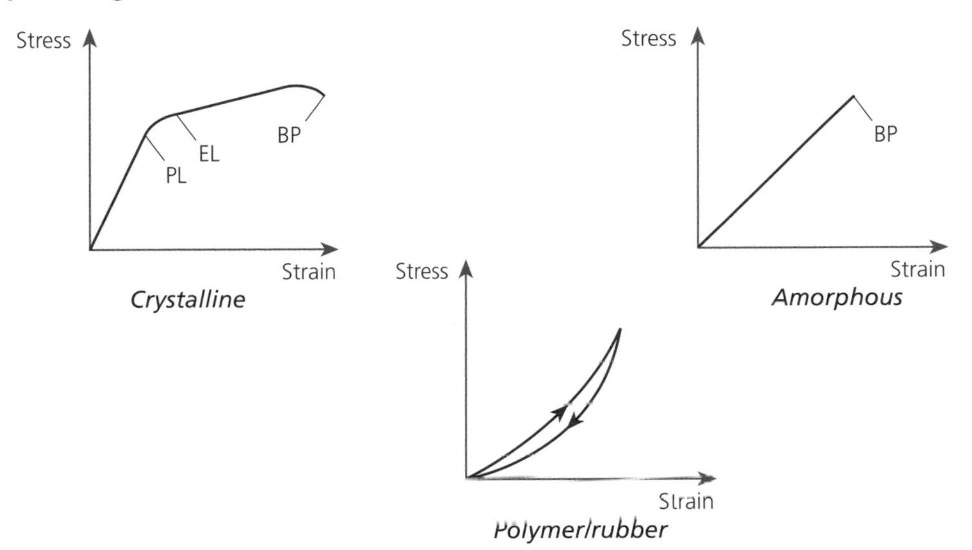

1.3 Work hardening

The hardness of crystalline materials depends on their structure and the number of grain boundaries that can hinder the flow of crystal planes. Work hardening by hammering reduces the size of the grains and introduces tangled dislocations.

Worked examples

Q1 A mass of 500 g is placed on the end of a nylon rope of diameter 1.4 cm. Calculate the stress applied.

$$\text{stress} = \frac{\text{force}}{\text{area}} = \frac{0.5 \times 9.81}{\pi \times (0.007)^2} = 32\,000\ \text{N m}^{-2}$$

Watch out for the units here.

Q2 An aluminium wire has an unstressed length of 2.5 m and a cross-sectional area of $1 \times 10^{-6}\ \text{m}^2$. The wire has a Young's modulus of 0.70×10^{11} Pa. What force must be applied to the ends to produce an extension of 5 mm?

$$\text{Young's modulus} = \frac{\text{stress}}{\text{strain}} \qquad \text{stress} = \frac{F}{1 \times 10^{-6}} \qquad \text{strain} = \frac{0.005}{2.5}$$

$$\therefore 0.70 \times 10^{11} = \frac{\dfrac{F}{1 \times 10^{-6}}}{\dfrac{0.005}{2.5}} \qquad \therefore F = 0.7 \times 10^{11} \times \frac{0.005}{2.5} \times 1 \times 10^{-6}$$

$$F = 140\ \text{N}$$

Q3 The length of a male femur bone is 43 cm and has a cross-section of 12 cm^2. What is the decrease in length in mm if the man lifts a bag of sand of mass 50 kg? The Young's modulus of bone under compression is 3.2×10^{10} N m^{-2}.

Half of the force on each leg.

$$\text{stress} = \frac{25 \times 9.81}{0.0012} \qquad \text{strain} = \frac{\Delta L}{0.43}$$

$$\therefore 3.2 \times 10^{10} = \frac{\frac{25 \times 9.81}{0.0012}}{\frac{\Delta L}{0.43}} \qquad \therefore \Delta L = \frac{25 \times 9.81 \times 0.43}{0.0012 \times 3.2 \times 10^{10}}$$

$= 2.75 \times 10^{-6}$ m
$= 2.8 \times 10^{-3}$ mm

You should now know:

- **the definitions of stress, strain and Young's modulus**
- **how to distinguish between limit of proportionality and elastic limit**
- **how to distinguish between elastic, polymeric and amorphous materials**
- **how to describe the structure of these materials and explain what happens when they are subjected to stress**

2 *Solids: force–extension graphs*

The stress–strain graph of an elastic material is a straight line for the application of small stresses. If we assume the cross-sectional area does not change, we can replace the stress–strain graph with a straight-line force–extension graph.

A simple spring obeys this graph.

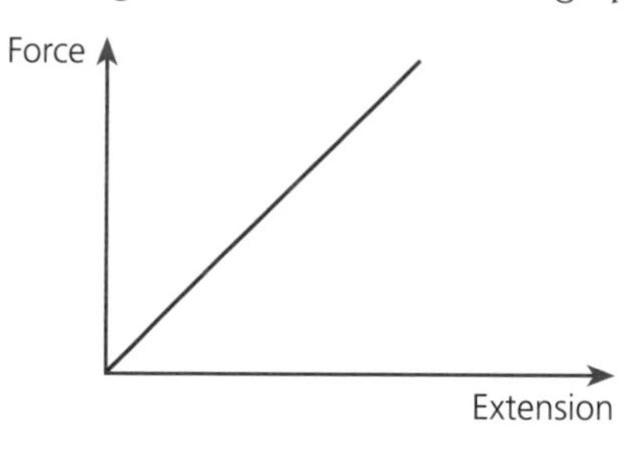

For this graph, force ∝ extension, or force = $k \times$extension. k is a constant, called the force or spring constant, which is the slope of the graph, and is measured in units of N m^{-1}.

2.1 Work done in stretching an elastic material

When a force moves through a distance, work is done on the elastic material, and the energy used is stored as elastic potential energy in the material.

If the graph of force against extension is not linear, the material has passed the limit of proportionality, but the work done is still the area under the graph.

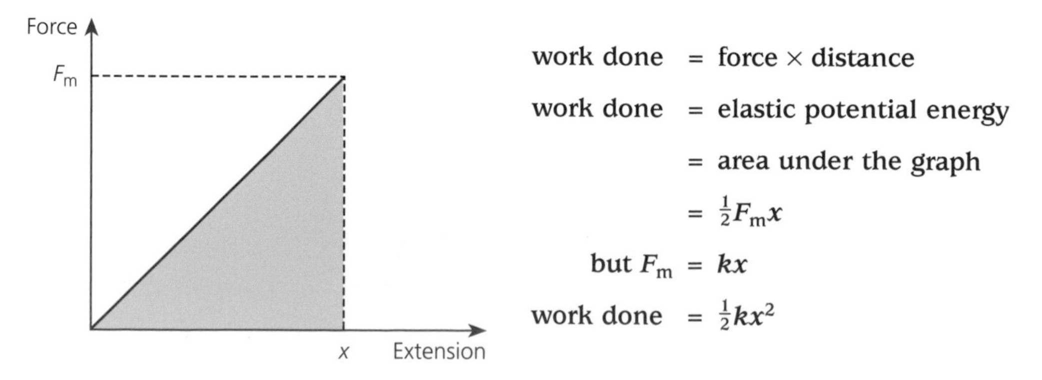

work done = force × distance

work done = elastic potential energy

= area under the graph

$= \frac{1}{2}F_m x$

but $F_m = kx$

work done $= \frac{1}{2}kx^2$

2.2 Force–extension graphs on an atomic scale

When the force applied is small, the atoms in a solid are displaced small distances from some equilibrium position, and forces are generated to restore the material to its original shape.

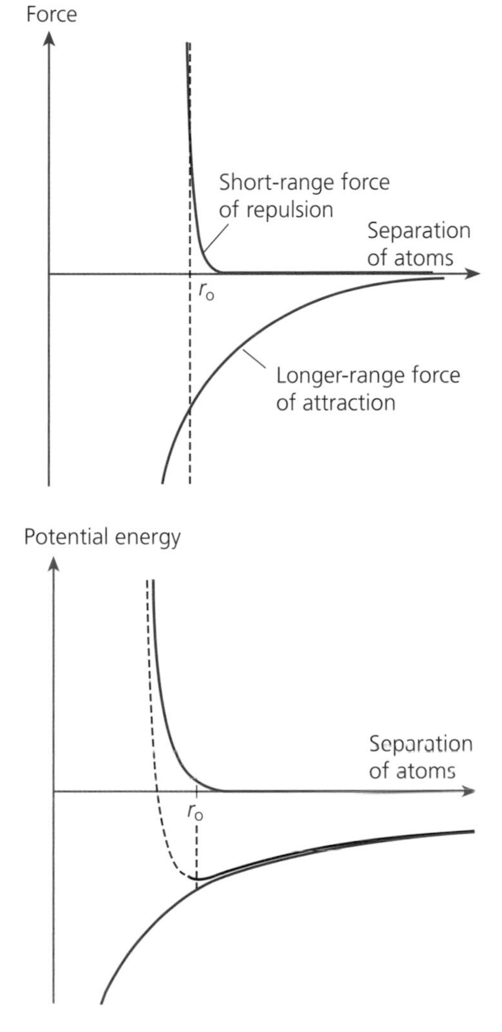

Restoring force under tension

A coulombic force is due to the charges on the atoms.

When atoms of the solid are pulled apart, an attractive coulombic force is generated, which attempts to pull the atoms back together.

Repulsive force under compression

When atoms of the solid are pushed together, a repulsive force is generated due to the overlap of electron orbits in the atoms. These two forces are equal and opposite at the equilibrium separation, r_0, of the atoms in the solid.

2.3 Potential energy graphs

When atoms are pulled together by coulombic forces, the potential energy reduces as work is done by the atoms as they move together.

As the atoms move together, a repulsive force is experienced, and work must be done on the atoms against the repulsive forces, so the potential energy increases.

The net potential energy is the sum of these two values and has a minimum at the equilibrium separation of the atoms in the solid.

Worked examples

$1 \text{ Pa} = 1 \text{ N m}^{-2}$
$1 \text{ GPa} = 1 \times 10^9 \text{ Pa}$

Q1 A copper wire of length 3.5 m and cross-section 1×10^{-6} m^2 is subject to a force of 45 N. Calculate the extension of the wire and the spring constant, assuming it obeys Hooke's law. Young's modulus for copper = 110 GPa.

$$\text{stress} = \frac{45}{1 \times 10^{-6}} \qquad \text{strain} = \frac{\Delta L}{3.5}$$

$$\therefore 110 \times 10^9 = \frac{\frac{45}{1 \times 10^{-6}}}{\frac{\Delta L}{3.5}} \qquad \therefore \Delta L = \frac{45 \times 3.5}{1 \times 10^{-6} \times 110 \times 10^9}$$

$$= 1.4 \times 10^{-3} \text{ m}$$

$$k = \frac{F}{x} = \frac{45}{1.4 \times 10^{-3}} = 3.5 \times 10^4 \text{ N m}^{-1}$$

Q2 A simple coil spring has a spring constant of 5.5 N m^{-1}. How much potential energy is stored in the spring when it is stretched through a distance of 7 cm?

$$\text{PE} = \text{work done} = \tfrac{1}{2}kx^2 = \tfrac{1}{2} \times 5.5 \times 0.07^2 = 1.3 \times 10^{-2} \text{ J}$$

Q3 When a metal is stretched past the elastic limit and the area under the stress–strain curve is measured, the elastic potential energy is found to have a value x joules. Why is all this potential energy not released when the stress is removed?

Energy is used in the plastic region to cause planes of atoms to slide past each other. This energy is not released when the force is removed.

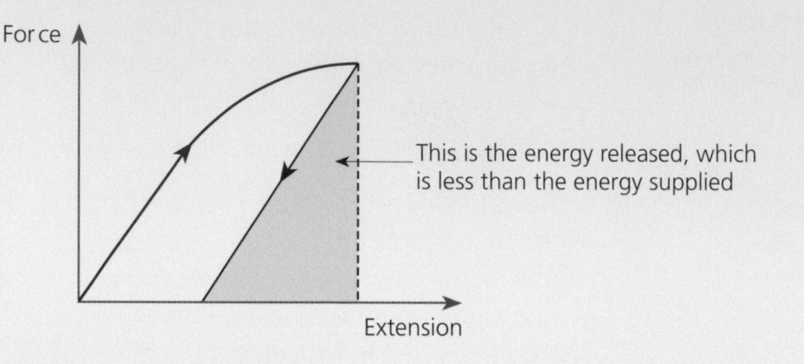

You should now know:

- **the definition of spring constant**
- **how to calculate the elastic potential energy when a material is stressed**
- **the shape of the potential energy curve for atoms in a solid and how the equilibrium separation is obtained from the graph**

3 *Fluids*

Fluid is a general term that describes both liquids and gases.

3.1 Common properties of liquids and gases

The property they have in common is that the atoms/molecules from which they are made are free to move randomly, and so the fluid is free to take up the shape of the container into which it is placed.

3.2 Differences between liquids and gases

The density of water at 4°C is greater than that of ice at 0°C.

The density of liquids is almost the same as that of solids, but the density of gases is much less than that of liquids or solids.

Atoms/molecules in liquids have little or no short-range order, but in a gas the atoms/molecules are at random positions.

See Topic 4, Section 4 on the kinetic theory of gases.

Liquid molecules are free to move around, but the motion of atoms/molecules in a gas is much more random, and the velocities of the particles are much higher than those in liquids.

3.3 Latent heat

It is useful to remember the values for water: LHF water = 344 kJ kg^{-1}; LHV water = 2.26 MJ kg^{-1}.

Energy must be supplied to convert a solid to a liquid and a liquid to a gas.

The specific latent heat of fusion is the energy per unit mass to turn a solid into a liquid.

The specific latent heat of vaporisation is the energy per unit mass to turn a liquid into a gas.

See Topic 4, Section 4.1 to remind yourself about the definition of pressure.

3.4 Pressure in a fluid

The pressure exerted by a column of fluid at the base can be calculated from the weight of fluid at the base.

The mass of a fluid is given by the density and the volume.

$$\text{mass of fluid} = \rho \times h \times A$$

$$\text{weight of fluid} = \text{force downwards} = \rho \times h \times A \times g$$

$$\textbf{pressure} = \frac{\textbf{force}}{\textbf{area}} = \rho h g$$

Note: the pressure depends only on the depth and the density of the fluid.

Atmospheric pressure

The air above the surface of the Earth generates a downwards pressure, which changes with time owing to changes in the density of the air, high and low pressures dictating weather patterns. Atmospheric pressure is measured using a mercury barometer by equating it to the pressure at the base of a column of mercury.

Standard atmospheric pressure is assigned a value of 0.760 m of mercury.

standard atmospheric pressure in pascals $= 0.76 \times 13.6 \times 10^3 \times 9.81$

$= 1.01 \times 10^5$ Pa

3.5 Buoyancy

The pressure in a fluid increases with depth. Hence, an object placed in a fluid will experience different pressures, depending on the depth.

The net force upwards is called the buoyancy force or the upthrust.

Consider a square block below the surface of a fluid. The pressure on the sides is equal and opposite and so cancels. Just considering the top and bottom surfaces

$$P_1 = \rho g d$$

$$P_2 = \rho g(d + h)$$

$$P_2 - P_1 = \rho g(d + h) - \rho g d$$

$$\text{force up} = \text{upthrust} = \rho g h A$$

$$= \text{weight of the fluid displaced}$$

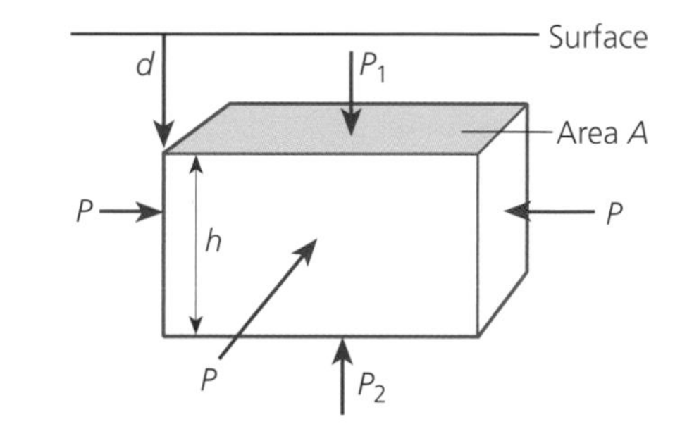

The upthrust equals the weight of the fluid displaced.

If the upthrust, which is equal to the weight of the fluid displaced, is greater than the weight of the object, the object will float; if less, the object will sink. This applies to any shape of object.

Archimedes' principle: **an object will float if it displaces its own weight of fluid.**

The centre of bouyancy is the centre of the displaced fluid.

Objects with a low centre of gravity are more stable when the object rolls to one side

You should be able to quote a value for the density of water from memory.

The absolute pressure would be this value plus atmospheric pressure.

Stability of floating objects

When an object rests in a fluid, the weight of the object acts down through the centre of gravity. The buoyancy force acts upwards through the centre of buoyancy. When the object rolls to one side, the position of the centre of gravity remains fixed but the centre of buoyancy moves. The object is stable (will right itself) if the centre of buoyancy moves to generate a restoring couple.

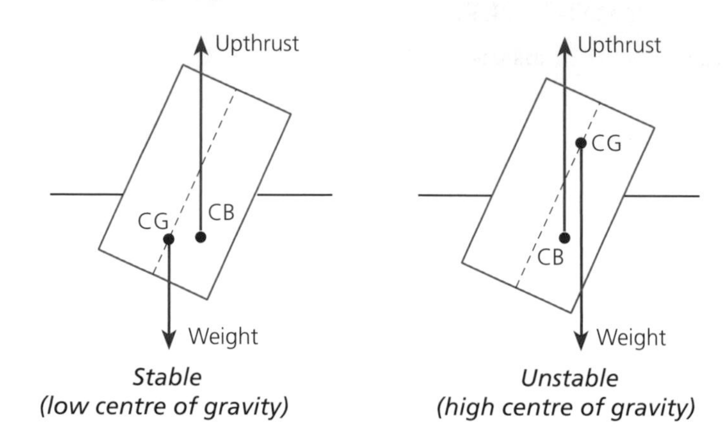

Stable (low centre of gravity) *Unstable (high centre of gravity)*

Worked examples

Q1 How much energy is required to produce one square ice cube with sides of 2 cm? Assume the water is at 0°C.

$$\text{volume} = (0.02)^3 = 8 \times 10^{-6}\ \text{m}^3$$

$$\text{mass of ice} = \rho \times 8 \times 10^{-6} = 1 \times 10^3 \times 8 \times 10^{-6} = 8 \times 10^{-3}\ \text{kg}$$

$$\text{energy} = 8 \times 10^{-3} \times 344 \times 10^3 = 2800\ \text{J}$$

Q2 What is the pressure at the bottom of a freshwater swimming pool of depth 2.5 m in pascal and mm of mercury?

$$\text{pressure} = \rho g h = 1 \times 10^3 \times 9.81 \times 2.5 = 2.5 \times 10^4\ \text{Pa}$$

$$2.5 \times 10^4 = 13.6 \times 10^3 \times 9.81 \times h$$

$$h = 0.18\ \text{m} = 180\ \text{mm}$$

Q3 An accurate set of scales measures the mass of a body to be 70 kg. The volume of the body is $6.8 \times 10^{-2}\ \text{m}^3$. If the density of air is $1.3\ \text{kg m}^{-3}$, what is the error in the mass, taking account of the buoyancy of the air?

$$\text{upthrust} = 1.3 \times 9.81 \times 6.8 \times 10^{-2} = 0.867\ \text{N}$$

$$\text{error} = \frac{0.867}{9.81} = 0.088\ \text{kg}$$

You should now know:

- **definitions of specific latent heat of vaporisation and fusion**
- **how pressure is related to depth in a fluid**
- **Archimedes' principle and the concept of buoyancy, and its application to stability**

4 Fluids in motion

In this section, we consider only the motion of a perfect liquid — that is one that does not offer any resistance to flow except through its own inertia. The liquid is considered to be incompressible.

4.1 Continuity equation

Consider fluid flowing down a tube with a velocity v.

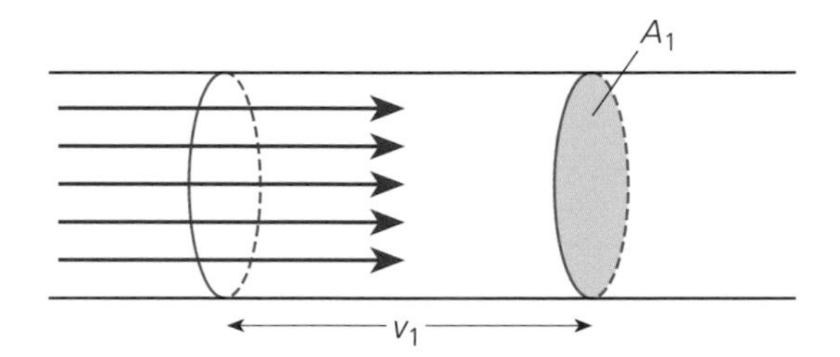

In 1 second, all the liquid in a tube of length v_1 will flow across the area A_1.

The volume of liquid flowing across an area A_1 in 1 second is A_1v_1.

If the area of cross-section of the tube changes, the volume flowing across the two areas/second must be the same, since the liquid cannot be compressed.

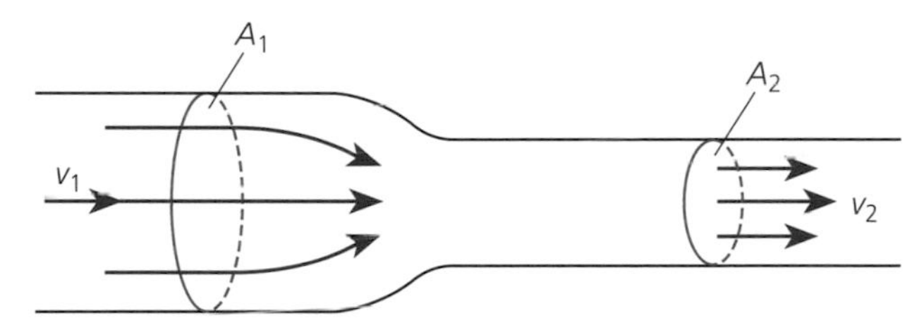

The continuity equation is $A_1v_1 = A_2v_2$ hence $v_2 > v_1$, since $A_2 < A_1$

4.2 Bernoulli's equation

When liquid flows down a horizontal tube and the area of cross-section changes, there is a change in the velocity of the fluid. If the velocity of the liquid increases, for a given mass m the kinetic energy of the liquid must also increase.

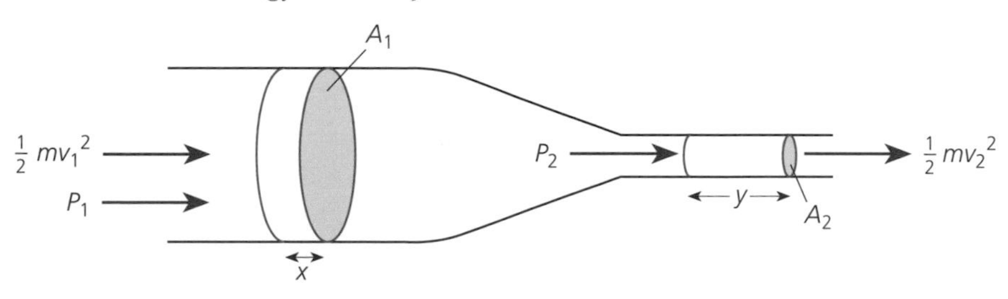

In this section, the mass m of liquid and the volume V are related by the density: $\rho = \frac{m}{V}$.

work = force × distance

pressure = $\frac{\text{force}}{\text{area}}$

You may only be required to understand the application of the Bernoulli equation to particular physical systems.

The law of conservation of energy must apply, and the increase in the kinetic energy is obtained from the work done on the liquid as it moves from one diameter tube to the other.

The work done in moving a volume of liquid V a distance x = force × x

In the large tube: work done = $P_1 \times A_1 \times x = P_1 \times V$

In the small tube: work done = $P_2 \times A_2 \times x = P_2 \times V$

Using the conservation of energy:

$$P_1 \times V + \tfrac{1}{2}mv_1^2 = P_2 \times V + \tfrac{1}{2}mv_2^2$$

$$P_1 - P_2 = \tfrac{1}{2}\frac{m}{V}(v_2^2 - v_1^2) = \tfrac{1}{2}\rho(v_2^2 - v_1^2)$$

This is **Bernoulli's equation.**

When a liquid flows down a tube that reduces in diameter, the velocity of the fluid increases and hence the pressure reduces.

4.3 Flowing gases

Much of the above can also be applied to flowing gases, the major difference being that gases are compressible and liquids are almost incompressible.

Applications of Bernoulli's theorem

The venturi meter

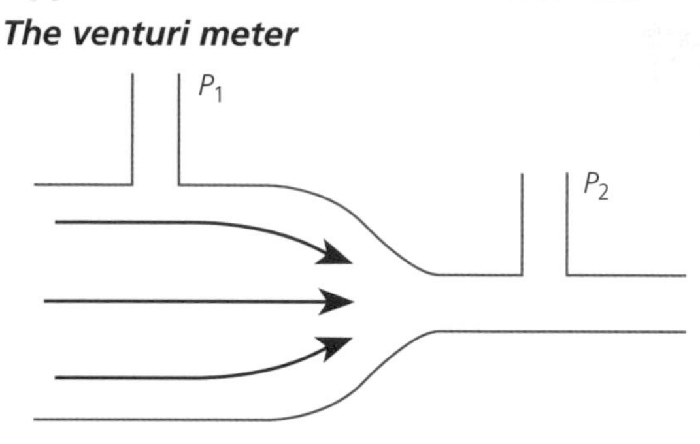

This is a device for measuring the flow of fluids by measuring the pressure difference between two pipes of different diameters.

flow rate $\propto \sqrt{P_1 - P_2}$

Application to aircraft wings

Aircraft wings are shaped so that the air flows a longer distance over the top than over the bottom. Hence, the air flows faster over the top than over the bottom. The pressure of the gas at the top is lower than the pressure of the gas below the wing. A net upward force is generated on the wing, which can be resolved into two components, the lift and the drag.

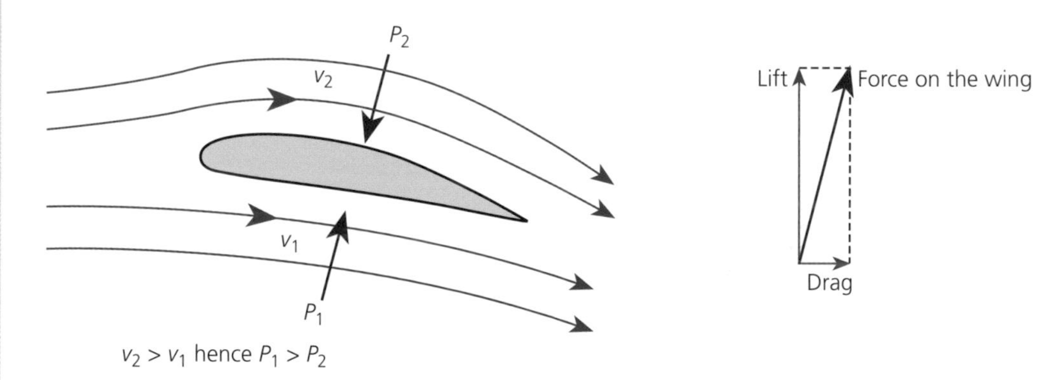

$v_2 > v_1$ hence $P_1 > P_2$

Refer to Topic 3, Section 1 to revise how to resolve a force into two components at right angles.

Application to sailing boats

The sail on a sailing boat is curved, and air at the front of the sail travels further than air at the back. A pressure difference is generated, producing a net force at right angles to the sail. When resolved into two components, it provides a force to drive the boat through the water and a sideways force balanced by an opposite force on the keel.

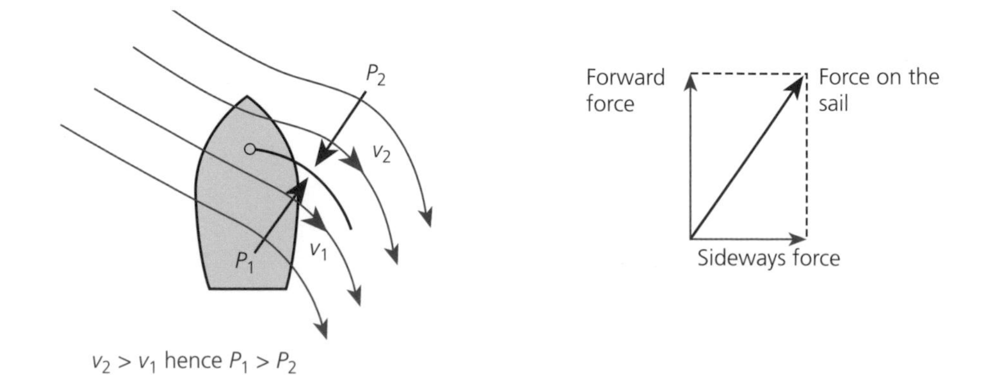

$v_2 > v_1$ hence $P_1 > P_2$

Remember to convert the diameter into an area

Worked examples

Q1 A fire hose consists of a pipe of diameter 5.8 cm, and at the outlet of the hose is a metal fitting with an exit diameter of 1.5 cm. If the velocity of water flowing down the hose is 5 m s^{-1}, what is the velocity with which the water leaves the exit?

$$v_1A_1 = v_2A_2$$

$$5 \times \pi \times \left(\frac{5.8}{2}\right)^2 = v_2 \times \pi \times \left(\frac{1.5}{2}\right)^2$$

$$v_2 = \frac{5 \times 5.8^2}{1.5^2} = 75\,\text{m s}^{-1}$$

Q2 In the previous example, what is the pressure difference across the outlet?

$$P_1 - P_2 = \tfrac{1}{2} \times 1000 \times (75^2 - 5^2)$$
$$= 500 \times (5625 - 25) = 2.8 \times 10^6\,\text{Pa}$$

Q3 A venturi meter, used to measure the flow of water, has an input diameter of 30 cm and a constriction diameter of 10 cm. If the velocity in the 30 cm diameter tube is 6 m s^{-1}, what pressure difference would be generated?

$$A_1v_1 = A_2v_2$$

$$\pi \times \left(\frac{30}{2}\right)^2 \times 6 = \pi \times \left(\frac{10}{2}\right)^2 \times v_2 \qquad v_2 = 54\,\text{m s}^{-1}$$

$$P_1 - P_2 = \tfrac{1}{2} \times 1000 \times (54^2 - 6^2)$$
$$= 500 \times (2916 - 36) = 1.4 \times 10^6\,\text{Pa}$$

You should now know:

- **the continuity equation**
- **how to explain in words the Bernoulli equation**
- **how to describe the application of the Bernoulli equation to the venturi meter, aircraft wings and sail boats**

TOPIC 6 Electricity

1 Current and potential difference

When an electric potential difference is placed across a material, an electric current flows.

1.1 Metals, semiconductors and insulators

Solids consist of arrangements of atoms bonded together by electrostatic forces. One method of classifying solids is based on their ability to conduct electricity. There are three major groups:

1. **Metals, which consist of an array of positive ions surrounded by free negative charges (electrons). Metals conduct electricity easily.**
2. **Semiconductors, in which the number of free electrons depends on temperature and impurities added to the solid. Semiconductors conduct electricity but not as well as metals.**
3. **Insulators, which contain few free electrons and do not conduct electricity well.**

The base unit in electrical measurements is the current. We define the basic unit of current as the ampere, the A in the MKSA (metre, kilogram, second, ampere) system of units.

1.2 Electric current

When electric charges flow down a metal wire, a current is said to flow.

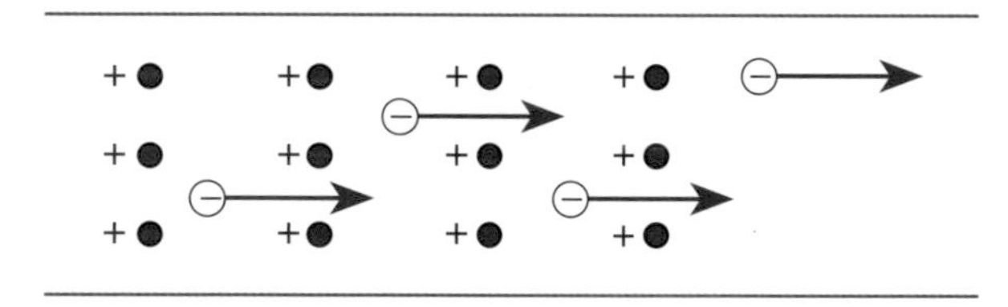

current (units, ampere) = rate of flow of charge = charge crossing an area of the wire per second

Current is used to define other electrical units.

When a current of 1 ampere flows in a wire, the rate of flow of charge is 1 coulomb/second (C s^{-1}).

Electric charges

The electric charges that flow down the wire are electrons, which carry a negative charge of 1.602×10^{-19} C. 1 ampere equals 6.24×10^{18} electrons s^{-1}.

Conventional current and electron current

When performing electrical calculations, the convention is to consider current flow in terms of positive electric charges.

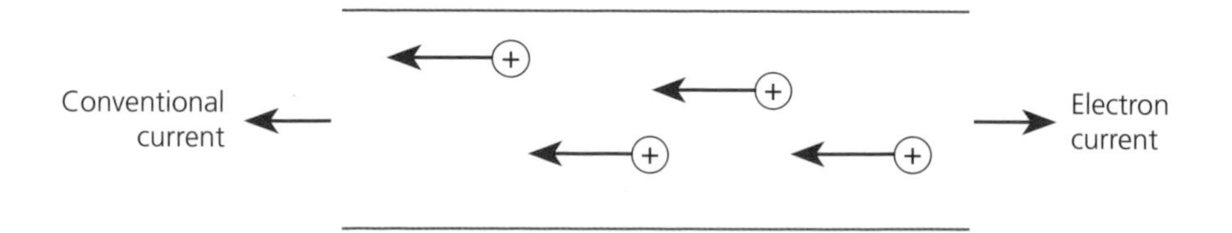

However, it must be remembered that electric current in metals is a flow of negative charges in the reverse direction.

Current density

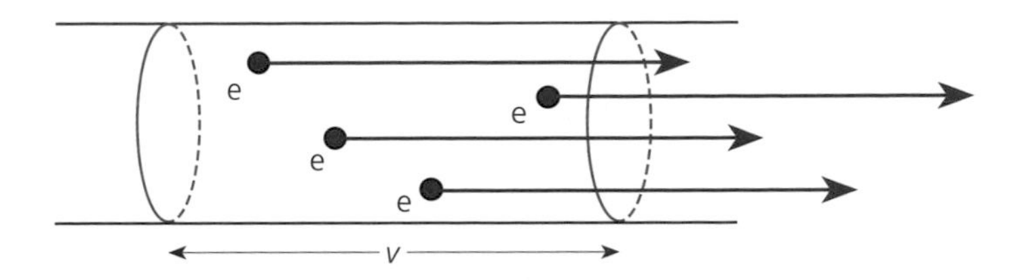

If charges of magnitude e coulombs flow down a wire with velocity v, then if the number of charges per unit volume is n, the total charge flowing across area A in 1 second is $n \times A \times v \times e$ = current.

$$I = nAve$$

The current density is defined as the current per unit area of cross-section.

$$J = \frac{I}{A} = \frac{nAve}{A} = nve$$

Current flow in liquids and gases

When a liquid or a gas contains charged particles, usually ionised atoms when positive ions and free electrons are present, then a current can flow. Charges of opposite sign move in opposite directions.

1.3 Potential difference

In order for a current to flow along a metal wire, and hence electric charges to move down the wire, energy is required. The potential difference indicates how much electrical energy is transferred when 1 coulomb of electric charge is moved from one point in the circuit to another.

The terms potential difference (p.d.) or voltage are often used to specify the potential difference.

The potential difference between two points in a circuit is defined as the energy converted from electrical energy to other forms of energy when 1 coulomb of charge moves between the two points.

Potential difference has units of joules/coulomb. 1 J/C = 1 V.

We can also define potential difference by considering the rate of flow of energy and hence the rate of flow of charge.

The potential difference between two points in a circuit is defined as the power (energy/second) converted from electrical energy to other forms of energy when 1 ampere (1 coulomb/second) flows between the two points.

Worked examples

Q1 A metal wire carries a current of 28 mA. Calculate the number of electrons crossing an area of the wire per second.

$$\text{current} = \frac{\text{charges}}{\text{time}} \qquad 28 \times 10^{-3} = \frac{\text{charges}}{1}$$

total charge $= 28 \times 10^{-3}$ C

$$\text{no. of electrons} = \frac{28 \times 10^{-3}}{1.6 \times 10^{-19}} = 1.8 \times 10^{17}$$

Q2 The density of free electrons in a metal is $1 \times 10^{28}\ \text{m}^{-3}$, and a wire of diameter 0.5 mm carries a current of 0.1 A. What is the velocity of electrons in the wire?

$$I = nAve$$

$$0.1 = 1 \times 10^{28} \times \pi \times \left(\frac{0.5 \times 10^{-3}}{2}\right)^2 \times v \times 1.6 \times 10^{-19}$$

$$v = 3.2 \times 10^{-4}\ \text{m s}^{-1}$$

Q3 An electric kettle is rated at 1 kW (1000 W) and is connected across a potential difference of 240 V. What current flows in the circuit? How long will it take the kettle to boil if it requires 2.4×10^5 J of energy to bring the water to boiling point?

$$\text{potential difference} = \frac{\text{power converted}}{\text{current}}$$

$$240 = \frac{1000}{I} \qquad \therefore I = 4.2\ \text{A}$$

$$\text{total time to boil} = \frac{2.4 \times 10^5}{1000} = 240\ \text{s} = 4\ \text{min}$$

You should now know:
- **the different current flows in metals, semiconductors and insulators**
- **that current is the rate of flow of charge; the units of current and charge**
- **the equations for current and current density in terms of charge/unit volume**
- **the two definitions of potential difference**

2 *Resistance and conductance*

It is important to note that the p.d. is the property that drives the current through the wire.

With a fixed p.d. across a solid, if the resistance of the solid increases, the current through the solid reduces.

In Section 1, we saw that when a p.d. is placed across a solid, a current flows and electrical energy is converted to other forms.

potential difference ∝ current

The size of the current, and so the energy converted, depends on the material across which the p.d. is placed. This is covered by introducing a constant into the above equation.

potential difference = resistance × current, or $V = RI$

This is Ohm's law.

resistance = potential difference/current, units volts/amps = ohms (Ω)

Conductance is simply the inverse of resistance.

conductance = current/potential difference, units Siemens (S)

2.1 Variation of resistance with temperature

The electrical resistance of a solid is due to collisions of the moving charges with the vibrating atomic lattice through which the charges flow. These vibrations increase with temperature, making the flow more difficult and increasing the electrical resistance.

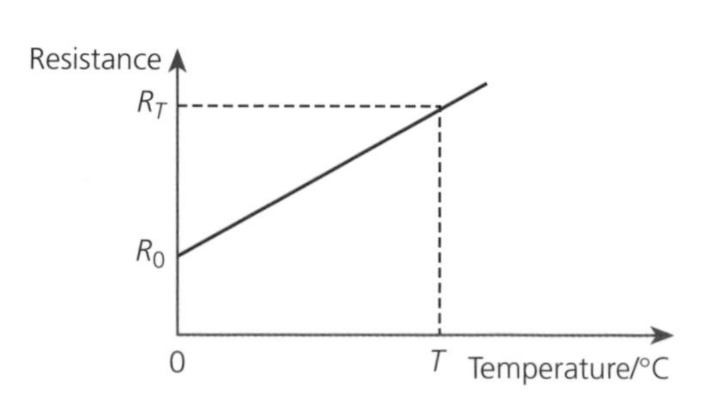

$$R_T = R_0(1 + \alpha T)$$

The temperature coefficient of resistance $\alpha = \frac{R_T - R_0}{R_0 T}$

Some solids — semiconductors for example — do not follow this general rule and the resistance decreases with increasing temperature.

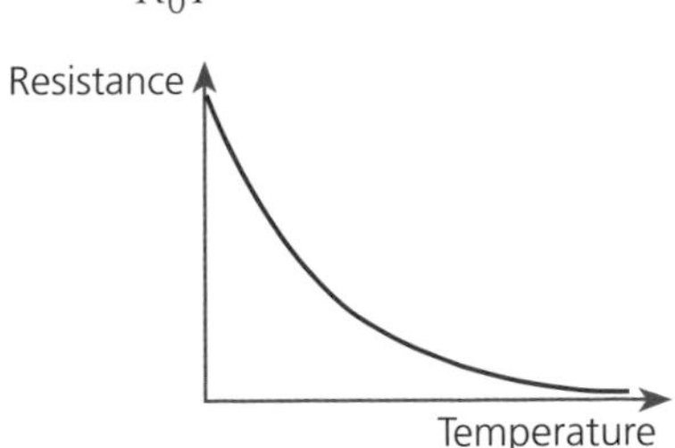

In a non-ohmic system, the resistance at any point is the ratio of voltage/current at the point concerned, *not* the slope of the graph at that point.

2.2 Resistance of other systems

When considering resistance, systems can be placed into one of two groups:

1 Ohmic systems, in which the resistance is a constant no matter what p.d. is applied.

2 Non-ohmic systems, in which the resistance changes as the voltage changes, e.g. diode, light bulb.

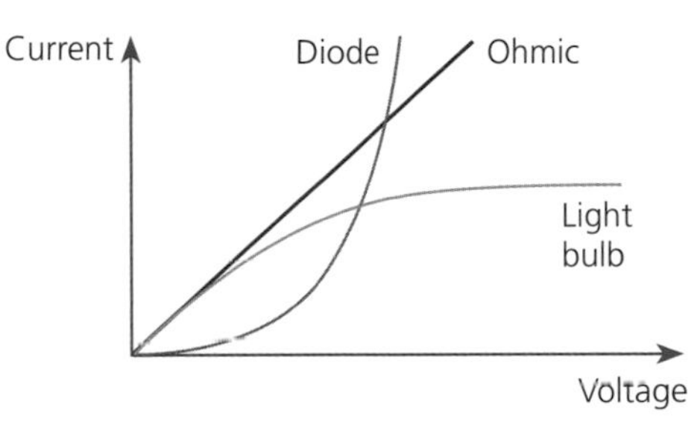

2.3 Resistivity and conductivity

The resistance of a solid depends on the dimensions of the solid and the material from which the solid is made.

$$\text{resistance} \propto \frac{\text{length}}{\text{area}}$$

This is converted into an equation with the inclusion of constant ρ.

Resistivity is the resistance across a unit cube of the material.

resistance $= \frac{\rho L}{A}$**, where** ρ **is the resistivity of the material (units, Ωm)**

or

conductance $= \frac{\sigma A}{L}$**, where** σ **is the conductivity of the material (units, S** m^{-1}**)**

Worked examples

Q1 What is the resistance of a 40 W and a 100 W light bulb when used with mains voltage with a p.d. of 240 V?

Using $V = \frac{\text{power}}{\text{current}}$, calculate the current through the bulb.

40 W: $I = \frac{40}{240} = 1.7 \times 10^{-1}$ A $\quad R = \frac{V}{I} = \frac{240}{1.7 \times 10^{-1}} = 1400\ \Omega$

100 W: $I = \frac{100}{240} = 4.2 \times 10^{-1}$ A $\quad R = \frac{V}{I} = \frac{240}{4.2 \times 10^{-1}} = 570\ \Omega$

Q2 A length of copper wire has a resistance of 3.4 Ω and is pulled so that its length doubles. What is the new resistance of the wire?

volume $= A \times L$ When pulled, the volume is the same.

$A' \times 2L = A \times L \quad \therefore A' = \frac{A}{2}$

$3.4 = \rho \frac{L}{A} \qquad R = \rho \frac{2L}{\frac{A}{2}} = \rho \frac{4L}{A} = 4 \times 3.4 = 14\ \Omega$

Q3 Two types of copper wire are used in houses: the diameter of one is 1.5 mm and of the other 2.5 mm. What would be the resistance of 10 m of each of these wires? The resistivity of copper = $1.7 \times 10^{-8}\ \Omega$ m at 20°C. The temperature falls to 10°C. What is the new resistance of the wires? The temperature coefficient of the resistance of copper = $0.004\ \text{K}^{-1}$.

(i) $R = \rho \frac{L}{A} = \frac{1.7 \times 10^{-8} \times 10}{\pi \times \left(\frac{1.5 \times 10^{-3}}{2}\right)^2} = 0.096\ \Omega$

(ii) $R = \frac{1.7 \times 10^{-8} \times 10}{\pi \times \left(\frac{2.5 \times 10^{-3}}{2}\right)^2} = 0.035\ \Omega$

$R_T = R_0(1 - \alpha T) \qquad R_{10} = 0.096(1 - 0.004 \times 10) = 0.092\ \Omega$

$R_{10} = 0.035(1 - 0.004 \times 10) = 0.034\ \Omega$

You should now know:

- **the definitions of resistance and conductance**
- **the variation of resistance with temperature for a metal and the definition of temperature coefficient of resistance**
- **the difference between ohmic and non-ohmic materials**
- **the definitions of resistivity and conductivity**

3 *Electromotive force*

In Section 1.3 potential difference was introduced in terms of the electrical energy converted to other forms of energy per unit charge. The law of conservation of energy states that this electrical energy must come from somewhere. In the case of electrical circuits, the energy is supplied by batteries, power supplies and generators to electrical charges, which flow around the circuit.

Note that e.m.f. has the same units as potential difference.

The electromotive force of the supply is the energy converted into electrical energy per unit charge supplied.

$$\text{e.m.f.} = \frac{\text{energy converted}}{\text{charge supplied}} \left(\frac{\text{units, joules}}{\text{coulombs}} = \text{volts}\right)$$

The term 'a 1.5 V battery' means that the e.m.f. is 1.5 V, and for a given source the e.m.f. is always constant.

All power supplies give out a specified e.m.f. For example, the small AAA batteries in MP3 players generate an e.m.f. of 1.5 V; the mains supply has an e.m.f. of 240 V.

3.1 Energy conversion

The energy supplied comes from many different sources, such as chemical energy in batteries or mechanical rotational energy in generators. The conversion of energy is never 100%, as some energy is lost in the conversion.

The energy gained by the charge is released as the charge travels around the electrical circuit. This energy is released in two places:

1 As the charge moves around the external electrical circuit — useful energy.

2 As the charge moves through the supply — wasted energy.

A perfect voltmeter will draw no current from the component across which it is placed when measuring the potential difference. Digital voltmeters draw very small currents.

If the flow of charge around the circuit is reduced to zero, no energy is lost in the supply and the energy supplied to the external circuit per unit charge will be a maximum and equal to the e.m.f.

The e.m.f. of a supply is the potential difference across the supply when no charges flow — the current is zero.

3.2 Internal resistance of the supply

As stated above, the energy supplied to a circuit is lost in two places — in the supply and in the external circuit. We can represent this with the following simple circuit.

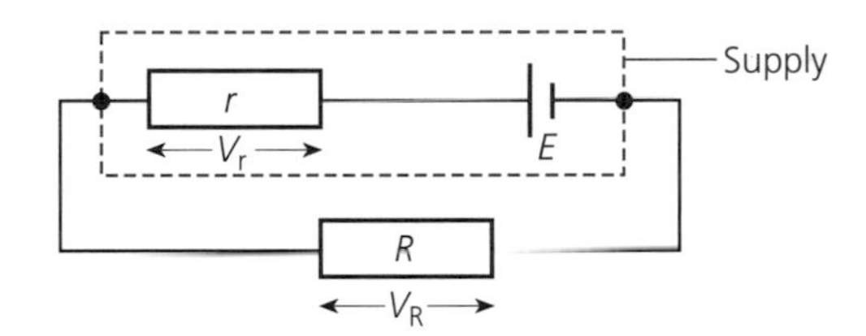

R is the total electrical resistance of the external circuit, and r is the resistance inside the supply (the internal resistance).

With this circuit, we can apply the law of conservation of energy and, for a unit charge flowing around the circuit, we write:

energy from the supply = energy converted in R + energy converted in r

e.m.f. = $V_R + V_r$, but $V_R = IR$ and $V_r = Ir$

e.m.f. = $IR + Ir = I(R + r)$

So a 12 V battery supplying current to a circuit will have less than 12 V across its terminals even though the e.m.f. is constant at 12 V.

In simple terms, the potential difference provided by the supply (the e.m.f.) must equal the sum of the potential differences across each of the components in the circuit, including the internal resistance of the supply.

V_R = e.m.f. − V_r

Real supplies

In many supplies, the internal resistance is very small compared with the external resistance and so the voltage across the internal resistance can be thought to be 0 and the e.m.f. = the voltage across the terminals.

3.3 Power in d.c. circuits

See Section 2.

Using the definition of potential difference, $V = \frac{\text{power converted}}{\text{current}} = \frac{P}{I}$, then power $P = VI$.

Using Ohm's law, $V = IR$, gives the power $P = I^2R$ and $P = \frac{V^2}{R}$.

When a resistance in a circuit carries a current I, the energy supplied to the resistance is all converted into heat. You should note that the heat is proportional to I^2 so, as the current increases, there is a rapid increase in the heat supplied.

Electrical power is transmitted over the grid and to railway engines by this method.

When large amounts of electrical power are required, it is important to reduce the current so that energy is not lost as heat in the wires. Using the equation power = IV, by increasing V we can reduce I and still supply the same amount of power.

The kilowatt hour (kWh)

Electricity suppliers calculate the cost of energy used in units of kWh.

This is the total energy supplied by a power source running at 1 kW for 1 hour.

$$1\,\text{kWh} = 1000\,\text{W} \times 3600\,\text{s} = 3\,600\,000\,\text{J}$$

Worked examples

Q1 A 9 V battery with an internal resistance of 0.1 Ω is connected to a 3 Ω resistance. What current flows across the circuit and what is the potential difference across the terminals of the battery?

$$E = I(R + r)$$

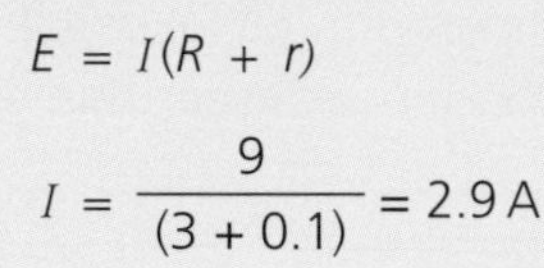

$$I = \frac{9}{(3 + 0.1)} = 2.9\,\text{A}$$

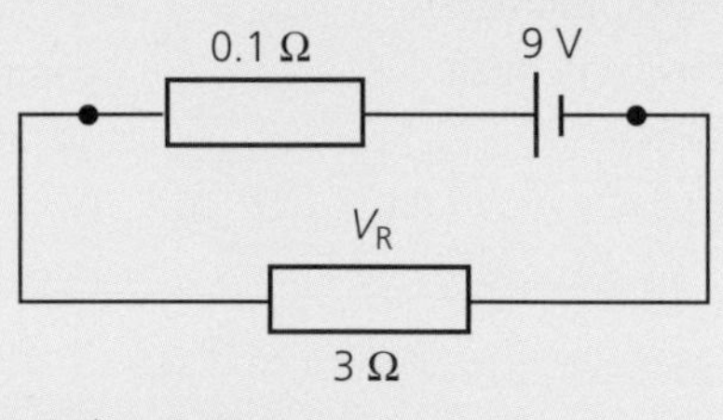

Potential difference across the battery = V_R = (9 – 0.29) = 8.7 V

Q2 A voltmeter of internal resistance 1×10^4 Ω is connected across a 12 V battery of internal resistance 0.2 Ω. What is the current in the circuit? What is the potential difference across the internal resistance of the battery?

$$E = I(R + r)$$

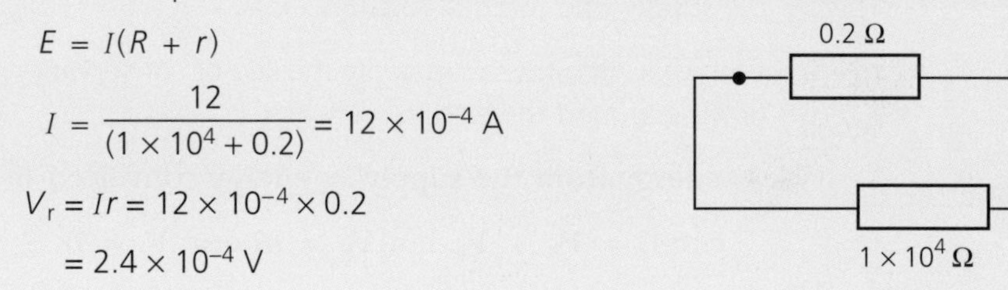

$$I = \frac{12}{(1 \times 10^4 + 0.2)} = 12 \times 10^{-4}\,\text{A}$$

$$V_r = Ir = 12 \times 10^{-4} \times 0.2$$

$$= 2.4 \times 10^{-4}\,\text{V}$$

Q3 An electric iron is rated at 600 W. What is the current drawn by the iron when connected to 240 V mains? How many units of electricity, kilowatt hours, are used when the total time the iron draws electricity is 40 min?

$$P = IV \qquad I = \frac{600}{240} = 2.5\,\text{A}$$

number of joules = 600 × 40 × 60 = 1 440 000 J

$$\text{number of kWh} = \frac{1\,440\,000}{60 \times 60 \times 1000} = 0.4\ \text{units}$$

or $\quad 0.6 \times \frac{40}{60} = 0.4$ units

You should now know:

- **the definition of e.m.f. and the difference between e.m.f. and p.d.**
- **the concept of the internal resistance of a supply**
- **the use of $E = I(R + r)$ for a simple circuit**
- **the equations for electrical power, $P = IV = I^2R = \frac{V^2}{R}$, and for the kilowatt hour**

4 Circuit calculations — Kirchhoff's laws

Electrical supplies drive charges around circuits, and we represent this in a simple circuit using currents. The connections to the supply dictate the direction of the charge flow and hence the direction of the current.

In circuit calculations, we use conventional currents, i.e. the flow of positive charges. Remember, in metals it is electrons moving in the reverse direction.

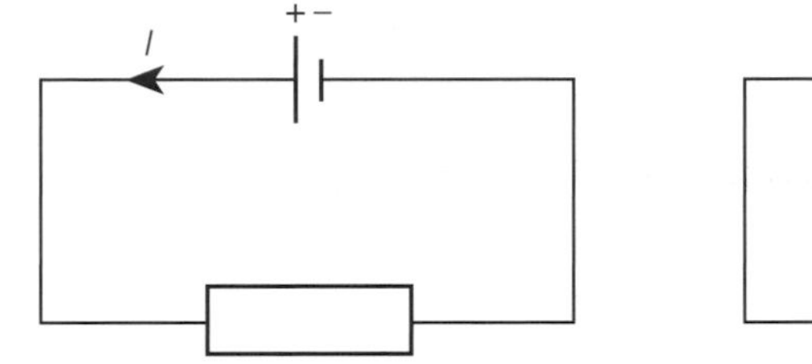

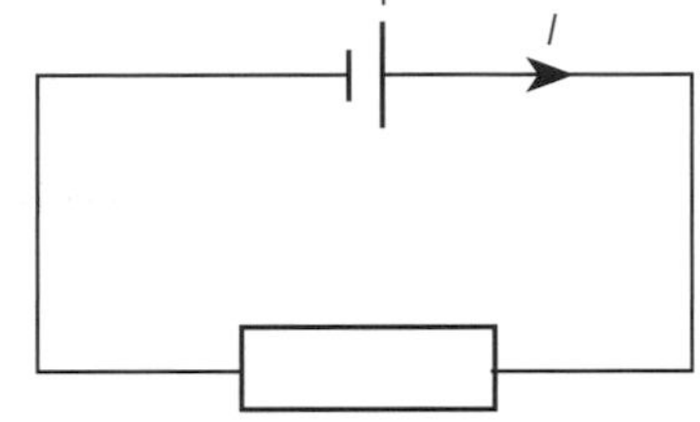

4.1 Dealing with larger circuits

In Section 3, we considered a simple electrical circuit with only one external component to the supply. Real electrical circuits consist of many components connected together. We need to introduce some simple rules to deal with these circuits.

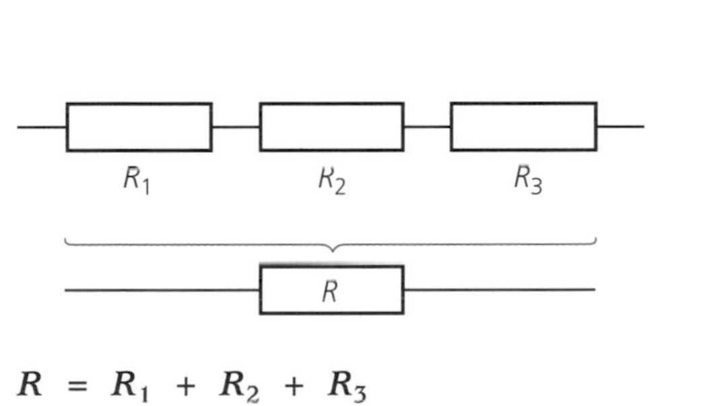

$$R = R_1 + R_2 + R_3$$

Resistances in parallel

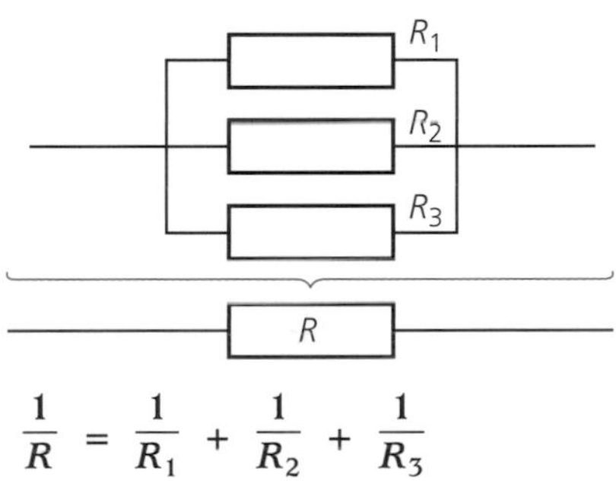

$$\frac{1}{R} = \frac{1}{R_1} + \frac{1}{R_2} + \frac{1}{R_3}$$

4.2 Dealing with complicated resistance networks

Resistance networks can be simplified by use of the above two equations in a suitable order as in the example below.

The simplification process requires a little thought as to how to progress.

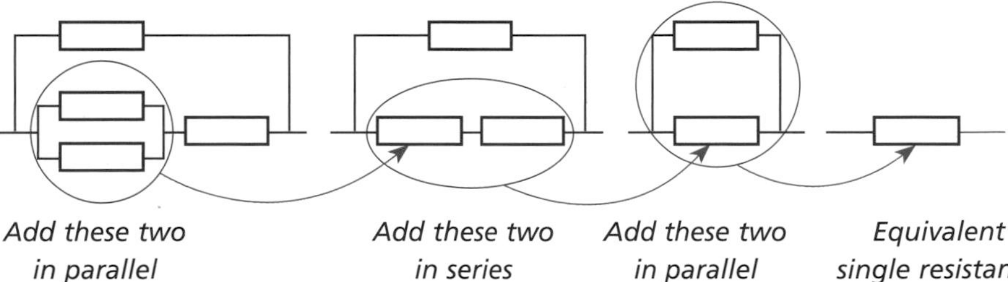

Add these two in parallel

Add these two in series

Add these two in parallel

Equivalent single resistance

4.3 The potential divider

The potential divider circuit consists of two resistances in series, placed across a potential difference. If the output current is zero, potential difference across R_2 is V_{out}.

This circuit is often used to produce a variable voltage output by adjusting R_2 using a variable

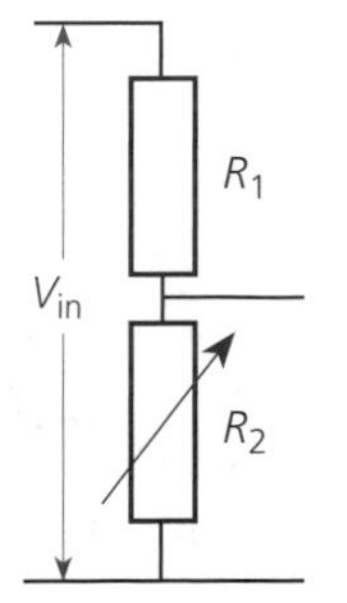

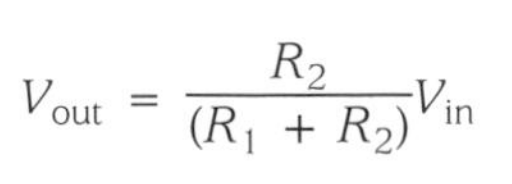

$$V_{out} = \frac{R_2}{(R_1 + R_2)} V_{in}$$

4.4 Kirchhoff's laws

In complicated circuits containing several supplies, circuit calculations using the simplification methods above are more difficult, and it is easier to make use of Kirchhoff's laws.

Kirchhoff's first law

Since electrical charge cannot be created or destroyed, when the charge flowing down a wire meets a junction of two or more wires the charge is divided between the junctions (formally stated in terms of current, rate of flow of charge).

The algebraic sum of the currents at any junction must be zero.

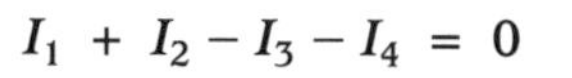

$$I_1 + I_2 - I_3 - I_4 = 0$$

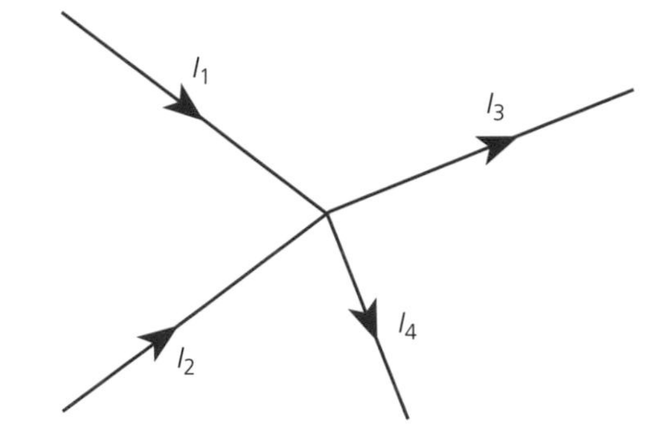

Remember e.m.f. is energy *in* per unit charge, and potential difference is energy *out* per unit charge.

Kirchhoff's second law

The second law is a statement of conservation of energy around a circuit.

Around any closed circuit, the algebraic sum of the e.m.f.s equals the algebraic sum of the potential differences.

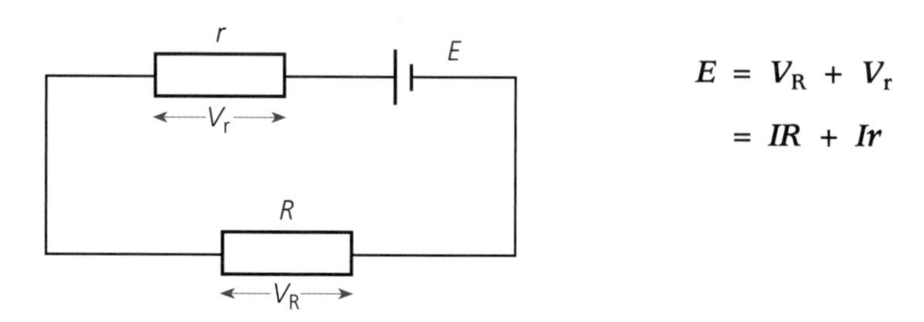

$$E = V_R + V_r$$
$$= IR + Ir$$

We can apply this to any circuit. Choose a direction for I and go round in the direction of I.

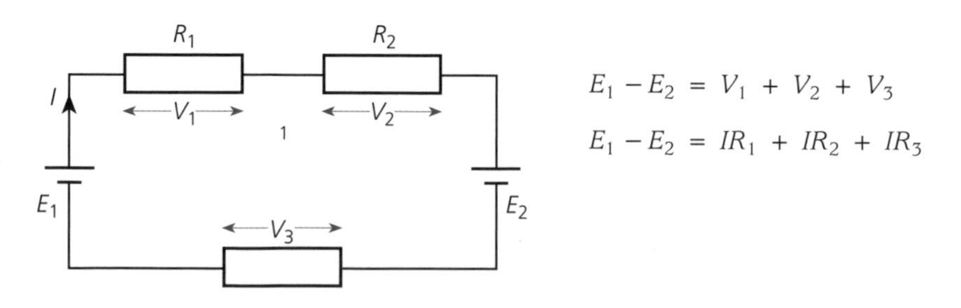

$$E_1 - E_2 = V_1 + V_2 + V_3$$
$$E_1 - E_2 = IR_1 + IR_2 + IR_3$$

If I is negative when calculated, the current must be in the reverse direction to the one chosen.

Worked examples

Q1 Calculate the equivalent single resistance of the following.

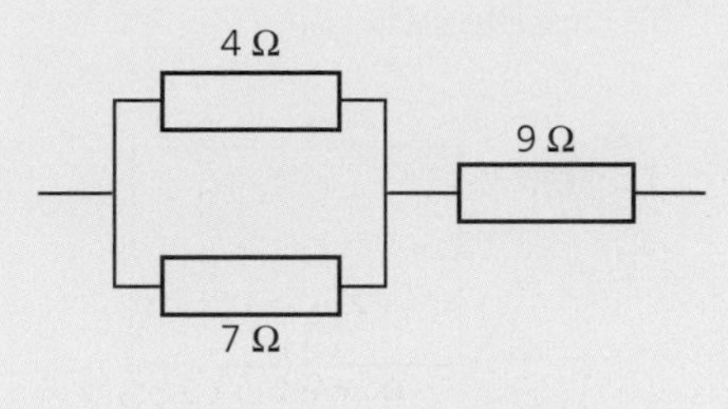

Calculate the parallel components first:

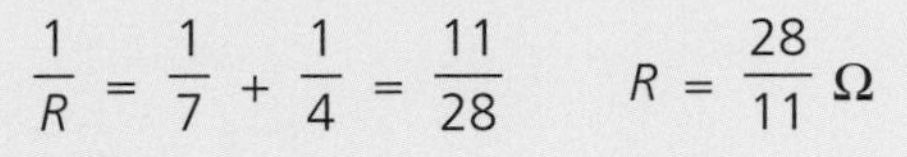

$$\frac{1}{R} = \frac{1}{7} + \frac{1}{4} = \frac{11}{28} \qquad R = \frac{28}{11}\ \Omega$$

Then the series:

$$R = 9 + \frac{28}{11} = \frac{127}{11} = 11.5\ \Omega$$

Q2 Calculate the current in the main section of the following circuit.

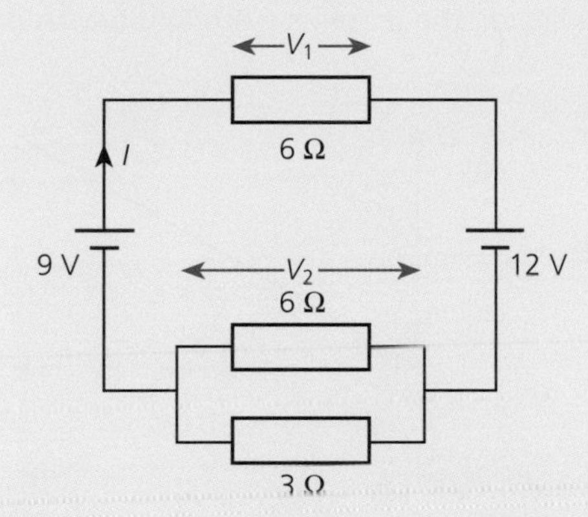

Calculate the parallel resistance:

$$\frac{1}{R} = \frac{1}{6} + \frac{1}{3} = \frac{3}{6} \qquad R = 2\ \Omega$$

Using Kirchhoff: $9 - 12 = V_1 + V_2$

$$-3 = 6I + 2I$$

$$I = -\frac{3}{8}\ \text{A}$$

in the reverse direction to that shown.

Q3 Using Kirchhoff's law, calculate the current through each of the parallel resistances in the circuit in question 2.

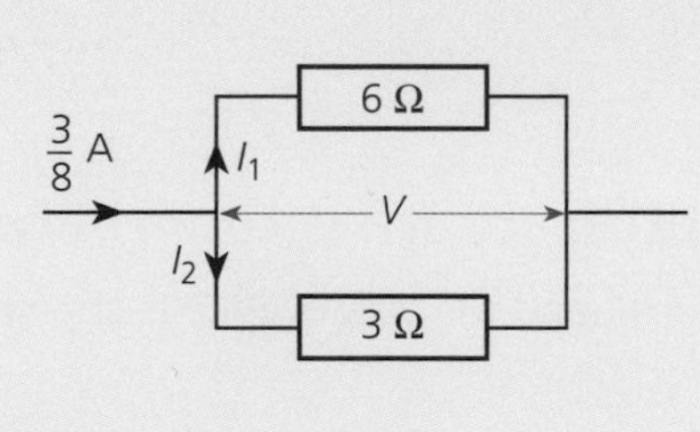

$V = 6I_1 \qquad V = 3I_2 \qquad I_2 = 2I_1$

But $I_1 + I_2 = \frac{3}{8}$

$$I_1 + 2I_1 = \frac{3}{8} \qquad I_1 = \frac{1}{8}\ \text{A}$$

$$I_2 = \frac{2}{8}\ \text{A}$$

You should now know:

- **the rules for the addition of resistors in series and parallel**
- **Kirchhoff's two laws and their application to simple circuits**

5 *Capacitance*

A capacitor is a circuit component that stores electrical charge. It consists of two metal plates placed close together and separated by an insulating layer called a dielectric. The insulating layer can be any suitable solid, liquid or gas.

5.1 Charging a capacitor

If a capacitor is placed across a source of e.m.f., a current will flow initially, transferring electrical charge from one plate to the other. During this process, the potential difference across the plates of the capacitor will rise.

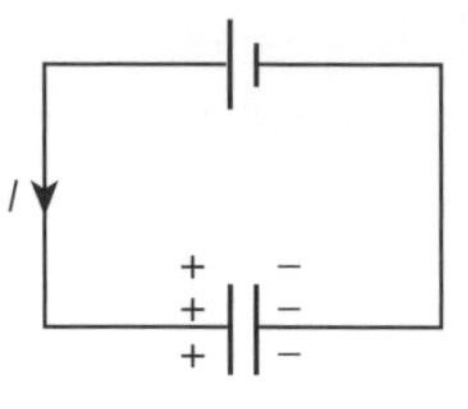

The amount of charge that can be stored depends on the size of the plates, the distance between the plates, and the material used as the dielectric.

The current flow will stop when the potential difference across the capacitor plates equals the e.m.f. of the source. The capacitor is said to be **charged**.

As the potential difference increases, more charge is stored on the plates. For a given construction of capacitor, the graph is a straight line and the slope is a constant called the **capacitance**.

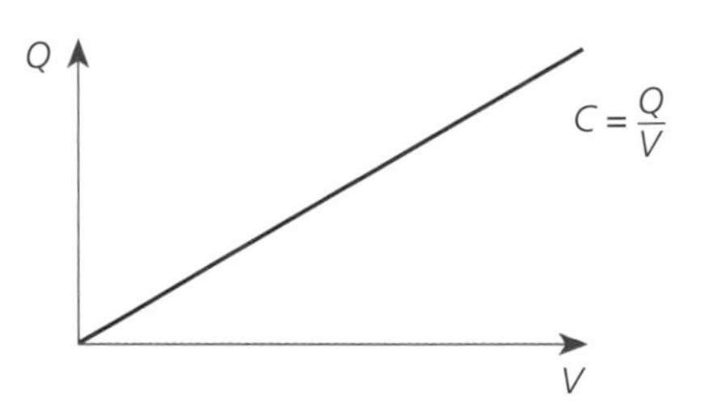

5.2 Units of capacitance

$$\text{capacitance} = \frac{\text{charge}}{\text{potential difference}} \qquad \text{units: } \frac{\text{coulombs}}{\text{volts}} = \text{farad (F)}$$

The farad is a very large unit for use in electrical circuits, and so smaller units are used:

- the microfarad, $1\ \mu F = 1 \times 10^{-6}$ F
- the nanofarad, 1 nF $= 1 \times 10^{-9}$ F
- the picofarad, 1 pF $= 1 \times 10^{-12}$ F

5.3 Parallel plate capacitor – vacuum between the plates

This is the simplest form, and the capacitance is shown to be:

$$C = \frac{\varepsilon_0 A}{d}$$

where A is the area of the plates and d is the separation, assuming a vacuum between the plates.

ε_0 is a constant called the permittivity of free space $= 8.9 \times 10^{-12}$ F m^{-1}.

5.4 Parallel plate capacitor – medium between the plates

In this case, the capacitance is:

$$C = \frac{\varepsilon_0 \varepsilon_r A}{d}$$

ε_r is the **relative permittivity** or the **dielectric constant**.

$$\varepsilon_r = \frac{\text{capacitance with a dielectric between the plates}}{\text{capacitance with a vacuum between the plates}}$$

The dielectric constant can range in value from 1 for a vacuum to over 300. Air has a dielectric constant of 1.0006.

Addition of capacitors

Capacitors in series

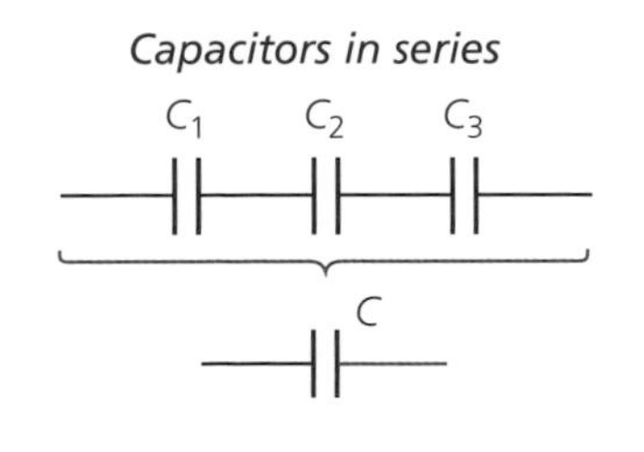

$$\frac{1}{C} = \frac{1}{C_1} + \frac{1}{C_2} + \frac{1}{C_3}$$

Capacitors in parallel

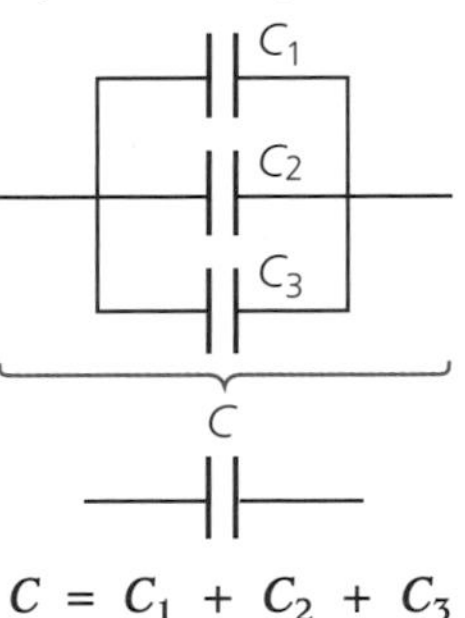

$$C = C_1 + C_2 + C_3$$

Note these equations are the reverse of those for resistance in series and parallel.

Series and parallel capacitors placed across a supply

The charge on each capacitor is the same

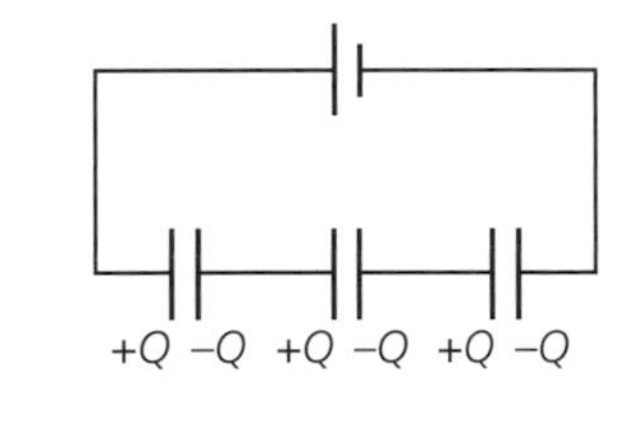

The voltage on each capacitor is the same

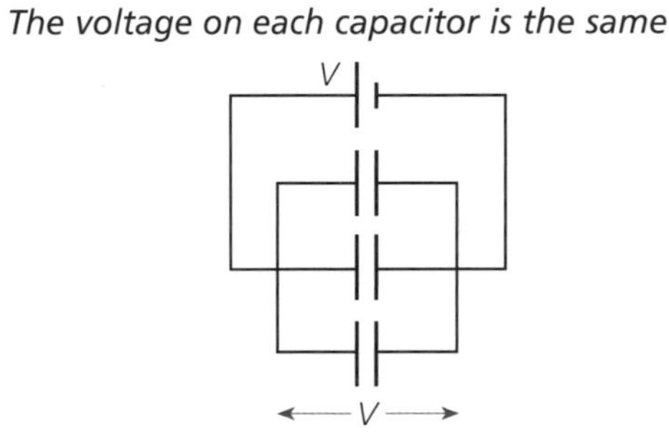

Worked examples

Q1 A simple capacitor consists of metal plates of area 0.04 m^2 in air, separated by a distance of 1 mm. What is the capacitance? What must be the separation of the plates for the capacitance to increase to 1 nF?

In this calculation $\varepsilon_r = 1$.

$$C = \frac{\varepsilon_0 \varepsilon_r A}{d} = \frac{8.9 \times 10^{-12} \times 0.04}{0.001} = 3.6 \times 10^{-10}\ \text{F} = 360\ \text{pF}$$

$$1 \times 10^{-9} = \frac{8.9 \times 10^{-12} \times 0.04}{d} \qquad d = \frac{8.9 \times 10^{-12} \times 0.04}{1 \times 10^{-9}} = 3.6 \times 10^{-4}\ \text{m}$$

Q2 A parallel plate capacitor is made from metal foil of dimensions 1 cm × 30 cm. It is separated by a dielectric of waxed paper, relative permittivity 2.7 and thickness 10^{-3} mm. What is the capacitance?

$$C = \frac{\varepsilon_0 \varepsilon_r A}{d} = \frac{8.9 \times 10^{-12} \times 2.7 \times 1 \times 10^{-2} \times 30 \times 10^{-2}}{1 \times 10^{-3}} = 72\ \text{pF}$$

Q3 What is the equivalent single capacitance of the following?

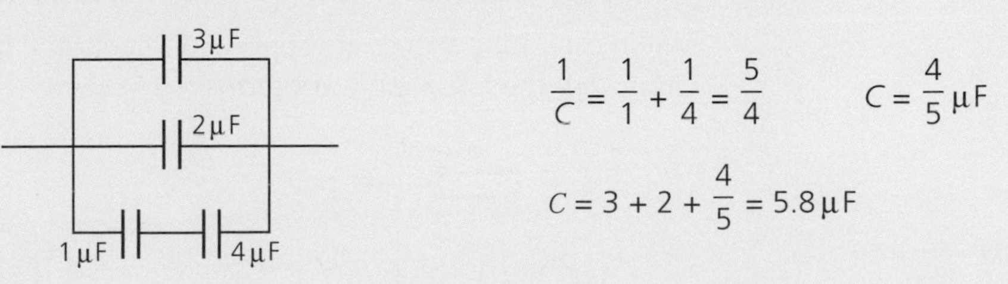

$$\frac{1}{C} = \frac{1}{1} + \frac{1}{4} = \frac{5}{4} \qquad C = \frac{4}{5}\ \mu\text{F}$$

$$C = 3 + 2 + \frac{4}{5} = 5.8\ \mu\text{F}$$

You should now know:

- **the equation of a parallel plate capacitor with and without a dielectric**
- **the use of the constants permittivity and relative permittivity**
- **the equivalent single capacitor of capacitors in series and parallel**

6 Charging and discharging a capacitor

When a capacitor is placed across a supply, it will charge up until the potential difference is equal to the e.m.f. of the supply. There will always be a resistance in the circuit, which, as a minimum, would be the internal resistance of the supply or an additional external resistance in the circuit.

These are exponential curves, and one of their characteristics is that they are asymptotic.

Charging

Discharging

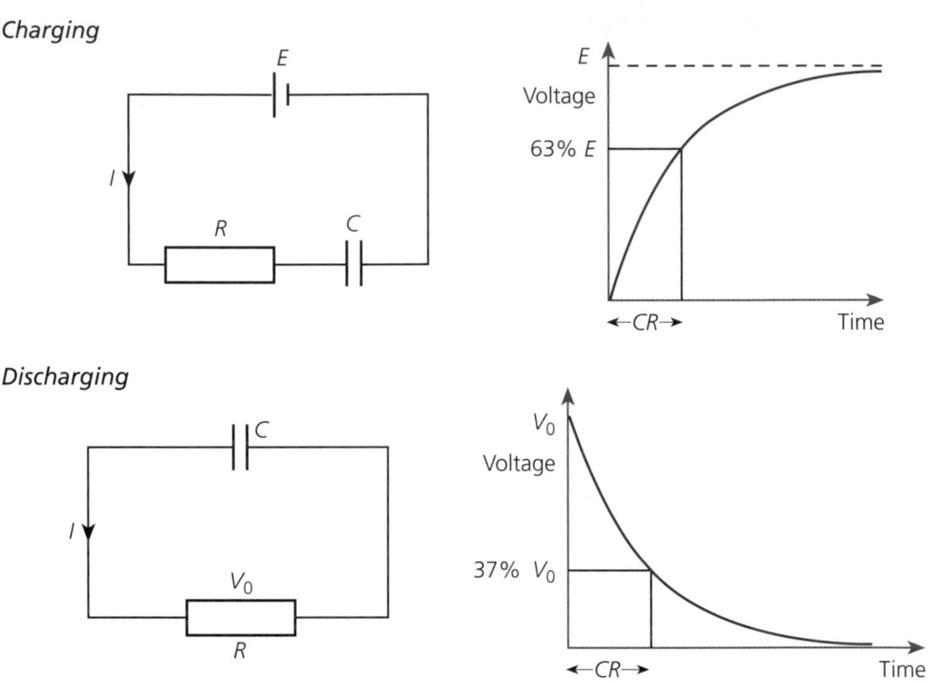

6.1 Equations for charging and discharging

You may not be required to know these equations, but check your specification. You will need to know the form of the charge and discharge curves.

Charging: $V = E(1 - e^{-\frac{t}{CR}})$

Discharging: $V = V_0 e^{-\frac{t}{CR}}$

CR is the **time constant** of the process.

Time constant

In a time of 1 time constant, the output reaches 63% of the final value, and in a time of 5 time constants, the output is >99% of the final value (it is within 1% of the final value).

6.2 Application of oscillating signals to a capacitor

When oscillating signals are applied to a capacitor, the capacitor charges and discharges, producing the following wave forms.

Capacitors are used in power supplies to smooth the rectified a.c. signal.

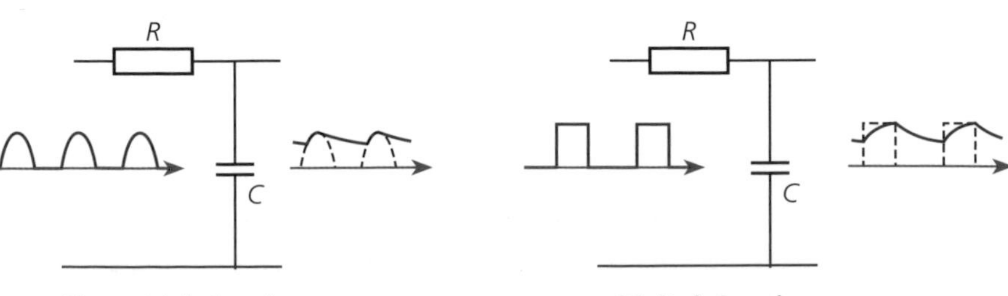

Sinusoidal signal *Digital signal*

The energy is stored in the electric field set up between the plates (see Topic 8, Section 4).

6.3 Energy stored in a capacitor

A capacitor is a store of electrical energy and if we remember that potential difference is the energy converted to other forms of energy per unit charge, we can calculate the energy stored in the capacitor.

As the capacitor charges, the potential difference across it increases. The energy stored is the average potential difference multiplied by the total charge stored.

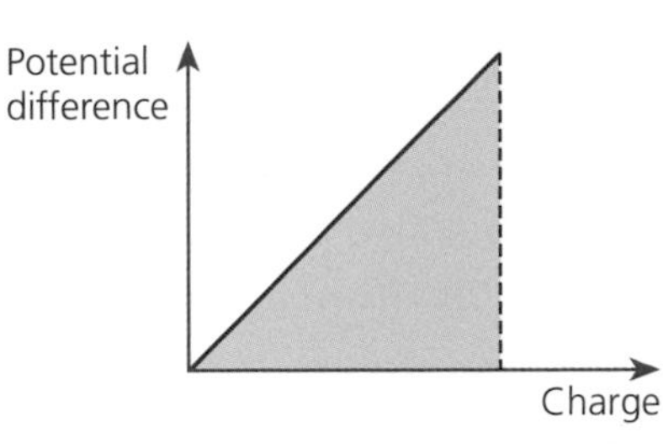

energy stored $= \frac{1}{2}V \times Q$

energy stored = area under the VQ graph

energy stored $= \frac{1}{2}VQ = \frac{1}{2}CV^2 = \frac{1}{2}\frac{Q^2}{C}$

Worked examples

Q1 A 440 μF capacitor is placed in a simple discharging circuit containing a resistance R. What value R will give a time constant of 0.1 s? How long will it take the voltage across the capacitor to fall to 1% of the initial value?

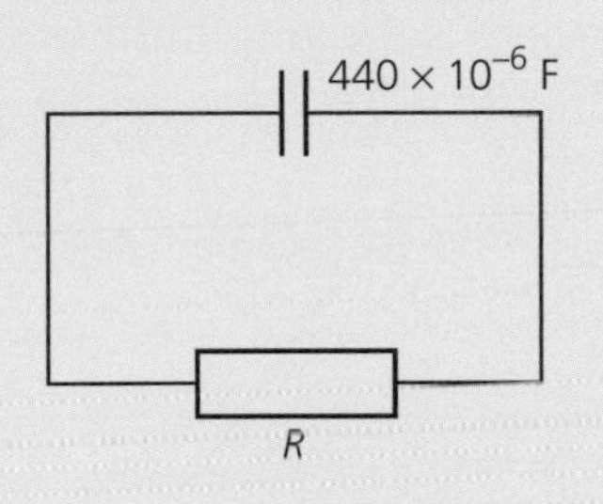

time constant = CR

$0.1 = 440 \times 10^{-6} \times R$

$R = 230\ \Omega$

For the output to fall to 1% of the initial value $= 5 \times CR = 0.5$ s

Q2 A 10 nF capacitor in series with a 1 MΩ resistance is placed across a 12 V supply. What is the voltage on the capacitor after a time of 5×10^{-3} s?

$$V = E(1 - e^{\frac{-t}{CR}})$$

$$= 12(1 - e^{\frac{(-5 \times 10^{-3})}{(10 \times 10^{-9} \times 1 \times 10^6)}})$$

$$= 12(1 - e^{-0.5}) = 12(1 - 0.606)$$

$$V = 4.7\text{ V}$$

Q3 A capacitor of 400 μF is placed across a 40 V supply. What is the charge on the plates and the energy stored? What voltage across the plates would reduce the energy stored to half the above value?

$$Q = CV = 400 \times 10^{-6} \times 40 = 0.016\text{ C}$$

$$\text{energy stored} = \tfrac{1}{2}CV^2 = \tfrac{1}{2} \times 400 \times 10^{-6} \times 40^2 = 0.32\text{ J}$$

$$\frac{0.32}{2} = \tfrac{1}{2} \times 400 \times 10^{-6} \times V^2 \qquad V^2 = 800$$

$$V = 28\text{ V}$$

You should now know:

- **the graphical form for the charge and discharge of a capacitor**
- **the definition of the term 'time constant'**
- **the three equations for the energy stored by a charged capacitor**

TOPIC 7 Waves and oscillations

1 *Oscillations and harmonic oscillations*

An oscillation is a motion in which an object undergoes a displacement that repeats itself after a fixed period of time. This is called a **periodic motion**. There are many simple examples of this in physics, such as vibrating strings, the pendulum on a clock, a mass vibrating on the end of a spring. There are also many occasions in real life where oscillations and their control are very important — for example, car suspension systems, loudspeakers etc.

1.1 Period and frequency

The period is the time taken for the oscillation to repeat itself (units, seconds).

One complete oscillation is called 1 **cycle** of the oscillating system.

The frequency of the oscillation is the number of times the oscillation is repeated in 1 second, or the number of cycles per second.

$$\text{frequency } (f) = \frac{1 \text{ second}}{\text{period}} = \frac{1}{T}$$

The units of frequency are **cycles per second** or **hertz (Hz)**.

The other parameter required is the **amplitude** of the oscillation, the parameter on the y-axis varying with time.

The displacement can be any physical parameter that varies with time: particle position, electric field etc.

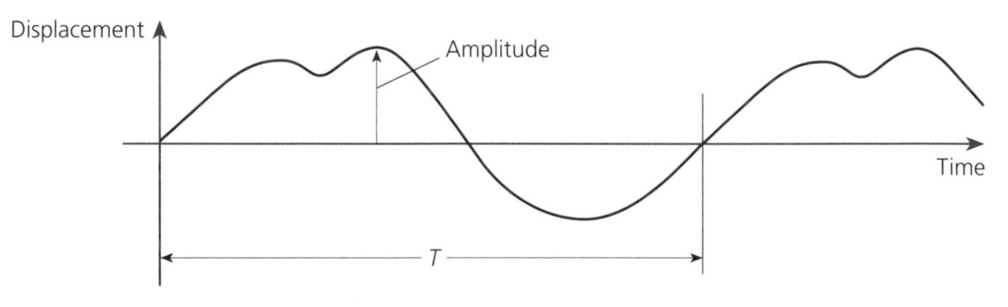

Examples of oscillations

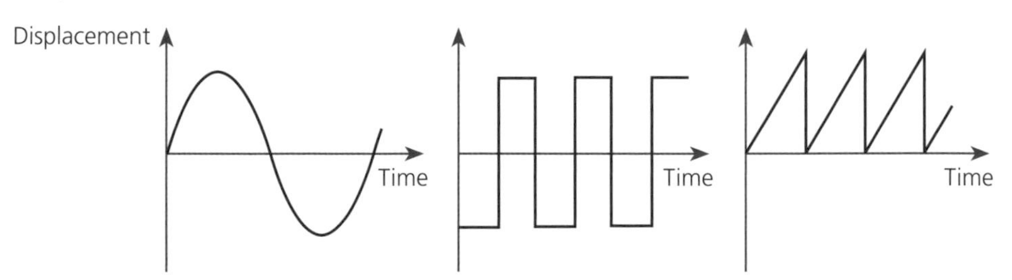

The shape of the displacement–time curve can vary in many different ways, as shown above.

1.2 Harmonic oscillations

Many oscillations fall into the largest group, called harmonic oscillations, whose displacements as a function of time vary according to sine or cosine functions.

Construction of harmonic oscillations

A harmonic oscillation can be constructed by considering a rotating arm, called a phasor, of size equal to the amplitude and rotating with a constant angular velocity, also called angular frequency.

Angular frequency

Angular frequency (ω) is the speed of rotation of the phasor measured in radians per second. As there are 2π radians in a circle, the time to make one rotation is T, the period.

$$T = \frac{2\pi}{\omega} \text{ but } T = \frac{1}{f}, \text{ hence } \omega = 2\pi f$$

To construct the harmonic oscillation, the x and y components of the phasor give the cosine and sine values of this oscillation as a function of time.

Note: in this case, the horizontal axis is plotted in terms of ωt, the angle turned through in radians, which increases with time.

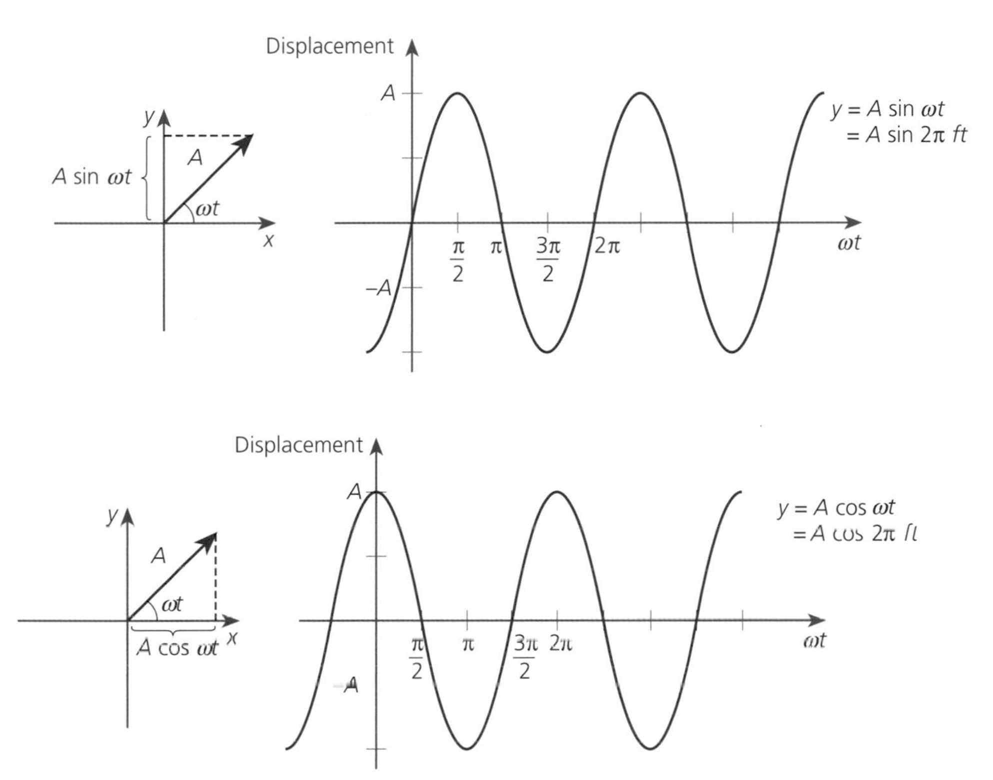

1.3 Phase

If we compare the two graphs for the sine and cosine waves, we can see that the sine wave lags behind the cosine wave by $\frac{\pi}{2}$ radians or $\frac{1}{4}$ wavelength.

See Section 4 on travelling waves for more information.

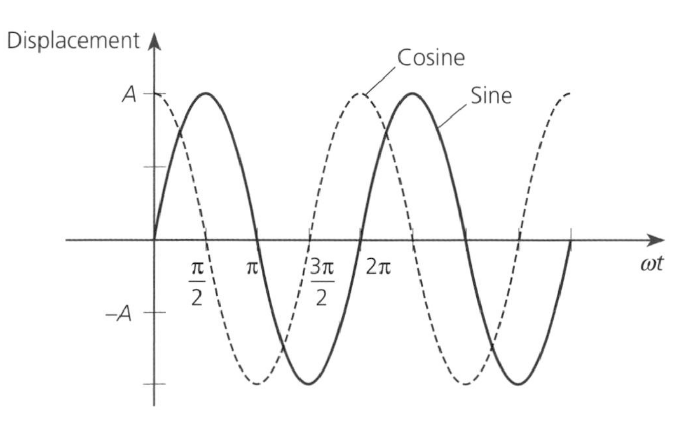

We say that there is a **phase difference** of $\frac{\pi}{2}$ between the sine and cosine waves.

Other phase differences

In many applications of harmonic waves, particularly in optics, we shall be considering waves out of phase by π or half a wavelength.

Worked examples

Q1 Calculate the period and frequency of an oscillation that has an angular frequency of 26 rad s^{-1}.

$$\omega = 26 \quad T = \frac{2\pi}{\omega} = \frac{2\pi}{26} = 0.24\,\text{s}$$

$$\text{frequency} = \frac{\omega}{2\pi} = \frac{26}{2\pi} = 4.1\,\text{Hz}$$

Take care here, as the angles are in radians, and you should make sure that your calculator is switched to radians when you calculate $\cos \omega t$.

Q2 A cosinusoidal oscillation has an amplitude of 3 and an angular frequency of 3 rad s^{-1}. What is the amplitude at a time of 20 s?

$$y = A \cos \omega t = 3 \cos 3 \times 20$$
$$= 3 \cos 60$$
$$= 3 \times -0.95 = -2.9$$

Q3 A harmonic wave B has a phase difference of $\frac{-3\pi}{2}$ with respect to a harmonic wave A. Sketch and label the two waves.

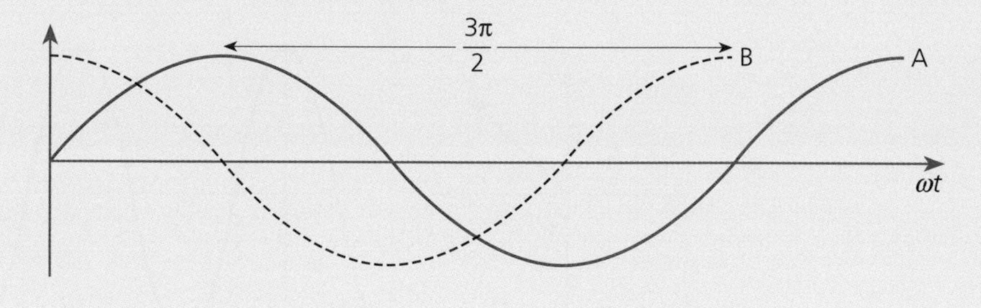

You should now know:

- **the definitions of period and frequency, and the linking equation**
- **the concept of harmonic oscillations and the equations for the amplitude of sine and cosine oscillations**
- **the idea of phase and phase difference**

2 *Simple harmonic motion*

In the previous section, we looked at oscillations in which the displacement of the particle followed a harmonic wave.

Note: the x-axis on this graph is plotted in terms of time.

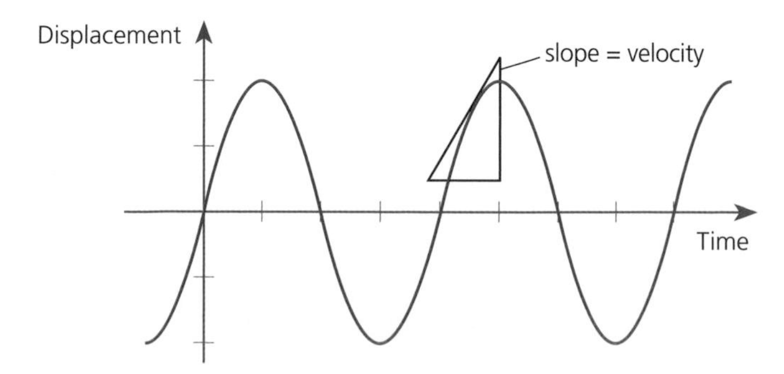

The slope of this graph will give the **velocity** of the particle, which is also a harmonic wave with a different phase, $\frac{-\pi}{2}$ with respect to the displacement curve.

You should see from this graph and the one below that when the displacement is zero, the velocity is a maximum and the acceleration is zero. When the displacement is a maximum, the velocity is zero and the acceleration is a maximum.

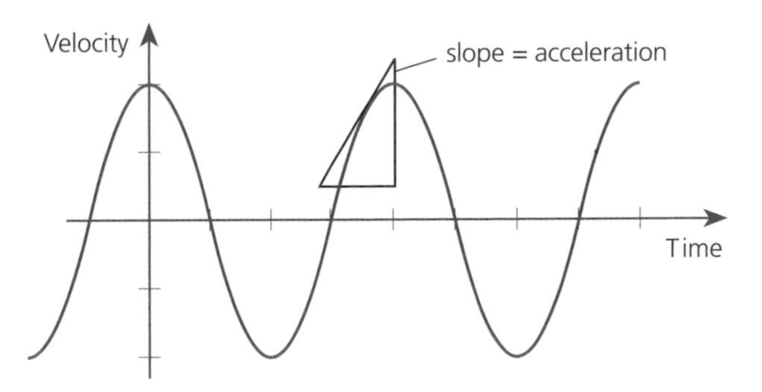

The slope of the velocity graph is the **acceleration**, which is also a harmonic wave with a phase of $-\pi$ with respect to the displacement curve. It is the negative of the displacement curve.

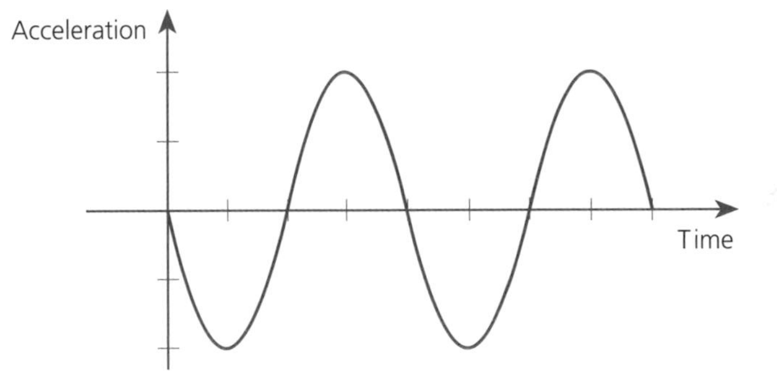

Hence, we can say that:

- acceleration $\propto$ −displacement
- acceleration $= -\omega^2$ displacement

An object that performs an oscillation such that acceleration = $-\omega^2$ × displacement is said to perform simple harmonic motion (s.h.m.).

2.1 Velocity

We can show by performing suitable calculations that:

The velocity at any displacement x is: $\mathbf{v} = \omega\sqrt{A^2 - x^2}$ where A is the amplitude

maximum velocity $\mathbf{v}_{max} = \omega A$

maximum acceleration $\mathbf{a}_{max} = -\omega^2 A$

$\mathbf{v}_{max}$ occurs when x = 0. See the above graphs.

2.2 Forces in oscillating systems

In this section we need to use vectors, since directions are important.

Consider a particle performing s.h.m. along the x-axis with an amplitude A.

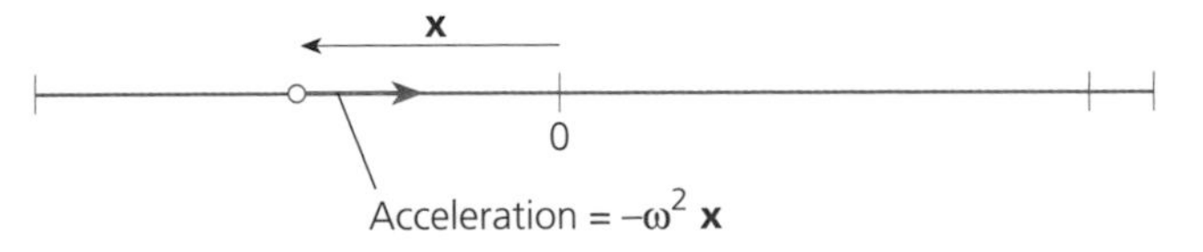

At maximum amplitude, the acceleration is a maximum and, since $\mathbf{a} = -\omega^2\mathbf{x}$, the acceleration is a maximum when **x** is a maximum. Since **x** and **a** are vectors, the acceleration is in the opposite direction to **x**, directed towards the centre.

The application of Newton's second law means that there must be a force **F** towards the centre of the oscillation, which is also proportional to −**x**.

2.3 Practical systems

Many practical systems perform oscillations in which a force proportional to the displacement from a central point is provided and is directed towards the central point. Such systems perform simple harmonic motion.

Mass on the end of a spring

When the mass is displaced downwards a distance $\mathbf{x}$ from the equilibrium position, a restoring force is generated.

Assuming Hooke's law, $\mathbf{F} \propto \mathbf{x}$.

The upward force, $\mathbf{F} = -k\mathbf{x}$, causes mass m to accelerate upwards.

Note that there is a negative sign because the force $\mathbf{F}$ is in the opposite direction to the displacement $\mathbf{x}$.

$$m\mathbf{a} = -k\mathbf{x}$$

$$\mathbf{a} = -\frac{k}{m}\mathbf{x} = -\omega^2\mathbf{x}$$

This is s.h.m. with $\omega = \sqrt{\frac{k}{m}}$

Simple pendulum

In this case, the force towards the centre

$$= mg \sin\theta = -\frac{mg\mathbf{x}}{L}$$

For small θ, $\sin\theta = \tan\theta = \frac{\mathbf{x}}{L}$.

The force towards the centre,

$$\mathbf{F} = -\frac{mg\mathbf{x}}{L}$$

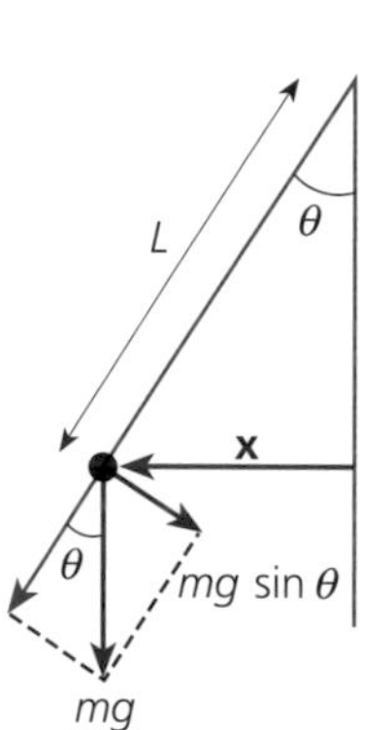

causes the mass to accelerate toward the centre.

$$m\mathbf{a} = -\frac{mg\mathbf{x}}{L} \qquad \mathbf{a} = -\frac{g\mathbf{x}}{L} = -\omega^2\mathbf{x}$$

This is s.h.m., with $\omega = \sqrt{\frac{g}{L}}$

Note that ω, and hence the period, is independent of the mass of the pendulum.

Worked examples

Q1 An oscillating system performs s.h.m. with an amplitude of 2 cm. If the period of the motion is 5 s, calculate the velocity of the particle at an amplitude of 0.5 cm. What is the maximum acceleration of the particle?

velocity $= \omega\sqrt{A^2 - x^2}$ $\qquad \omega = \frac{2\pi}{T} = \frac{2\pi}{5} = 1.3 \text{ rad s}^{-1}$

$= 1.3\sqrt{0.02^2 - 0.005^2}$

$= 1.3\sqrt{4 \times 10^{-4} - 2.5 \times 10^{-5}}$

$= 1.3\sqrt{3.75 \times 10^{-4}}$ $\qquad = 2.5 \times 10^{-2} \text{ m s}^{-1}$

maximum acceleration $= \omega^2 A = 1.3^2 \times 0.02 = 3.4 \times 10^{-2} \text{ m s}^{-2}$

Q2 A mass m on the end of a spring produces a force of 5.6 N and extends the spring a distance of 0.3 cm. If displaced further, so that oscillation takes place, what is the period of the oscillation?

Find k using $F = kx$:

$5.6 = k\, 0.003$

$k = 1870 \text{ N m}^{-1}$

$\omega = \sqrt{\frac{k}{m}}$ $\qquad T = \frac{2\pi}{\omega} = 2\pi\sqrt{\frac{m}{k}} = 2\pi\sqrt{\frac{5.6}{9.81 \times 1870}} = 0.11 \text{ s}$

Q3 What is the length of a simple pendulum on a clock so that the period of the oscillation is 1 s?

$$T = 2\pi\sqrt{\frac{L}{g}} \qquad 1 = 2\pi\sqrt{\frac{L}{g}}$$

$$\therefore L = \frac{1 \times g}{4\pi^2} = 0.25\ \text{m}$$

You should now know:

- **the relationship between displacement, velocity and acceleration in an oscillating system**
- **the defining equation for simple harmonic motion**
- **the application of s.h.m. to simple systems**

3 Energy in simple harmonic motion

An oscillating system contains kinetic energy and potential energy.

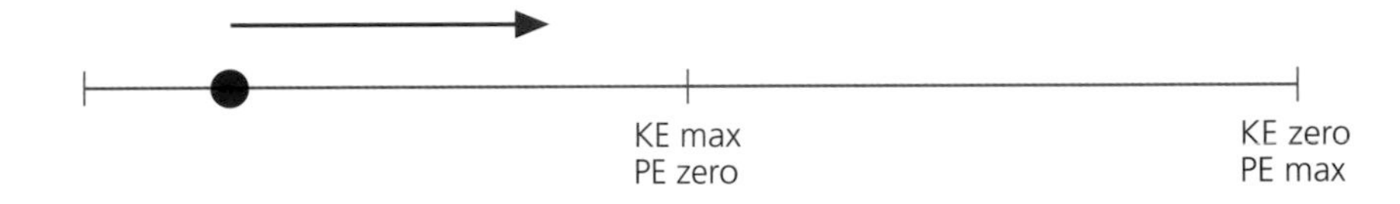

During an oscillation, there is a transfer of energy between these two forms, as shown above.

total energy at any point in the oscillation = kinetic energy + potential energy

$v = \omega\sqrt{A^2 - x^2}$. See Section 2.1.

The **kinetic energy** = $\frac{1}{2}mv^2 = \frac{1}{2}m\omega^2(A^2 - x^2)$, but the total energy equals the maximum kinetic energy when $x = 0$.

kinetic energy$_{\textbf{max}}$ = $\boldsymbol{\frac{1}{2}m\omega^2A^2}$

So the **potential energy** = $\boldsymbol{\frac{1}{2}m\omega^2A^2 - \frac{1}{2}m\omega^2(A^2 - x^2) = \frac{1}{2}m\omega^2x^2}$.

Potential energy is stored in many different forms, depending on the system being considered:

- a pendulum stores gravitational potential energy
- a vibrating mass on a spring stores potential energy in the spring, and gravitational potential energy if the spring vibrates vertically
- for vibrating molecules, potential energy is stored as electrostatic potential energy

3.1 Decay of s.h.m.

In real systems, the total energy is not constant. Energy is lost in many ways, depending on the system concerned, but mainly owing to frictional forces.

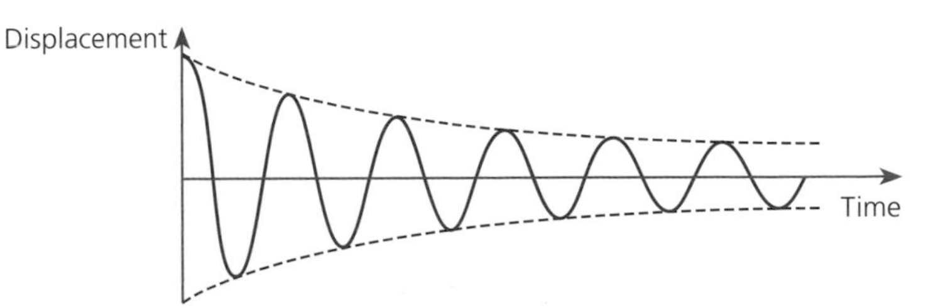

It can be seen that the amplitude of the oscillation decays away over a period of time, and the oscillations are said to be **damped**.

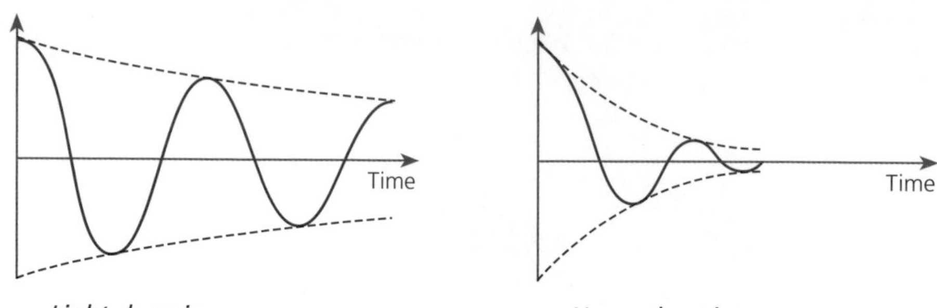

3.2 Forced vibrations

If the oscillations in real systems are to continue, extra energy must be added to the system to replace the energy lost. Such vibrating systems have forced vibrations.

All oscillating systems have a **natural frequency** of vibration, which is easily found by displacing the system from its equilibrium position and allowing it to vibrate.

3.3 Resonance curves

In these curves, the amplitude of the driver is kept constant. f_n is the natural frequency of oscillation of the system.

These are a set of curves that plot the ratio of the amplitude of the driven oscillation to that of the driver, called the **amplitude ratio**, as a function of driver frequency.

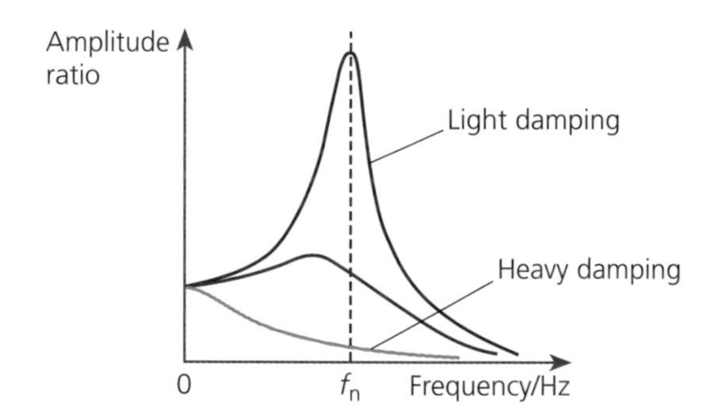

In the real world, there are occasions when resonance is to be avoided in systems that can oscillate, such as suspension bridges, the suspension systems on cars, trains etc. On other occasions, resonance is required, as in the design of oscillators in electronic circuits, for example.

Note the following from the graph:

- With light damping, when the driven frequency is close to the natural frequency, the oscillations produced have a large amplitude. This is called resonance.
- As the damping increases, the size of the resonance peak reduces and the position of the peak shifts to lower frequencies.
- With heavy damping, no resonant peak is observed.

Worked examples

Q1 A mass on a spring performs vertical oscillations. Show that the total energy is given by $\frac{1}{2}kA^2$, where k is the spring constant.

total energy of an oscillating system = $\frac{1}{2}m\omega^2A^2$

$$\omega = \sqrt{\frac{k}{m}} \text{ for a mass on a spring}$$

$$\text{total energy} = \frac{1}{2}m\frac{k}{m}A^2 = \frac{1}{2}kA^2$$

Q2 A simple pendulum consists of a 0.5 kg mass on the end of a massless string 1 m long. Calculate the period of the oscillation and the maximum velocity during the swing if the mass rises through a maximum height of 0.001 m.

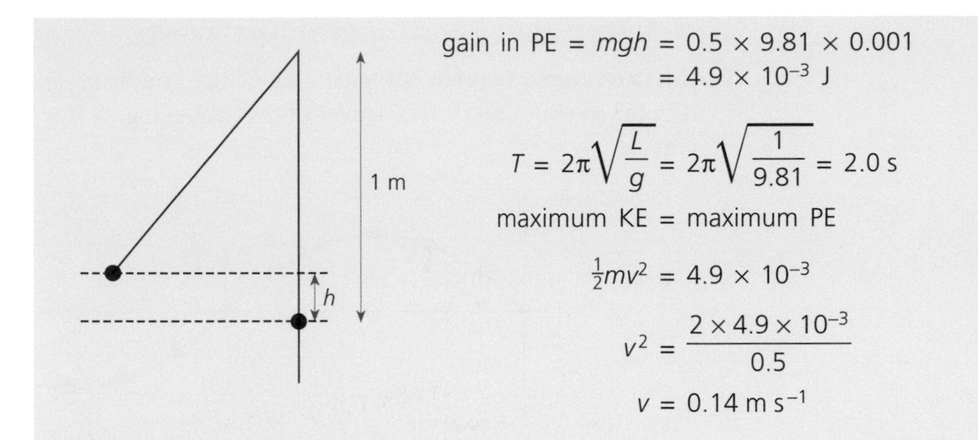

gain in PE = mgh = $0.5 \times 9.81 \times 0.001$
$= 4.9 \times 10^{-3}$ J

$$T = 2\pi\sqrt{\frac{L}{g}} = 2\pi\sqrt{\frac{1}{9.81}} = 2.0\text{ s}$$

maximum KE = maximum PE

$$\tfrac{1}{2}mv^2 = 4.9 \times 10^{-3}$$

$$v^2 = \frac{2 \times 4.9 \times 10^{-3}}{0.5}$$

$$v = 0.14\text{ m s}^{-1}$$

Q3 A hydrogen molecule consists of two hydrogen atoms of mass 1.67×10^{-27} kg, as shown below.

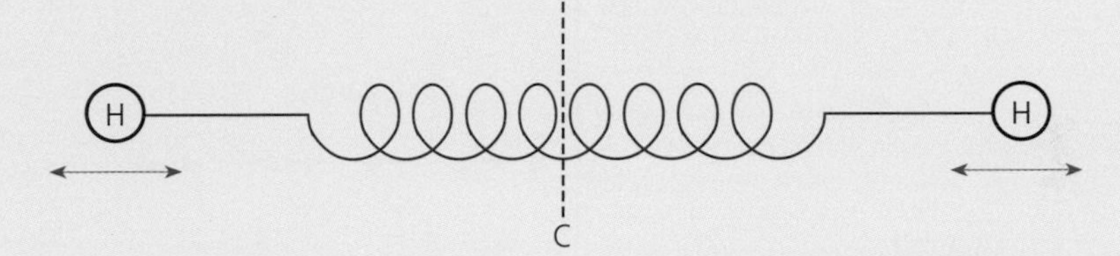

The spring constant for each atom is 1.1×10^3 N m^{-1}, and the total vibration energy is 1.3×10^{-19} J. Calculate the amplitude and frequency of the vibration.

Consider just one atom:

total energy = $\tfrac{1}{2}kA^2 = \dfrac{1.3 \times 10^{-19}}{2}$ J (for half the system)

$0.65 \times 10^{-19} = \tfrac{1}{2} \times 1.1 \times 10^3 \times A^2$ $\quad A = 1.1 \times 10^{-11}$ m

$$\omega = \sqrt{\frac{k}{m}} \qquad f = \frac{1}{2\pi}\sqrt{\frac{k}{m}} = \frac{1}{2\pi}\sqrt{\frac{1.1 \times 10^3}{1.67 \times 10^{-27}}} = 1.3 \times 10^{14}\text{ Hz}$$

You should now know:

- **the difference between free and forced vibrations**
- **the equations for the kinetic energy of an oscillating system**
- **the concept of resonance and how to explain the term 'natural frequency of oscillation'**

4 *Waves*

A wave is a disturbance that propagates from one place to another. The disturbance can be in the form of the oscillation of particles if the wave travels through a medium (as in the case of water or sound waves), or it can be an oscillation in electric and magnetic fields (as in the case of electromagnetic waves travelling through space).

In Section 6, we consider standing waves — where energy is not propagated.

Some wave motions require a medium through which to propagate, whereas electromagnetic waves can travel through a vacuum.

Waves that transfer energy from one place to another and are called **progressive waves**.

4.1 Transverse and longitudinal waves

In transverse waves, the particles of the medium or the fields oscillate perpendicular to the direction of travel of the wave, e.g. water waves and electromagnetic waves.

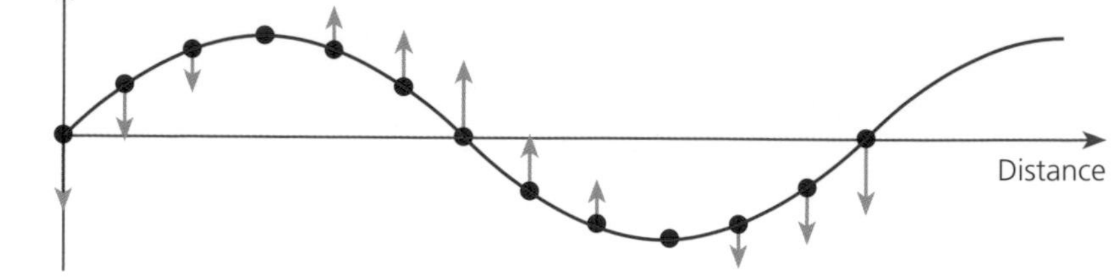

In longitudinal waves, the particles oscillate along the direction of travel of the wave, e.g. sound waves in air or water.

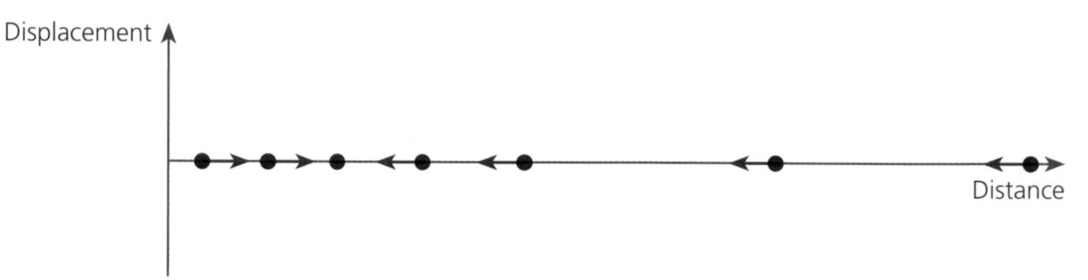

Look back at Section 1 on harmonic oscillations to remind yourself of the details.

4.2 Harmonic waves

Consider a harmonic wave in which the displacement of the particles of the medium are plotted as a function of distance at one instant of time.

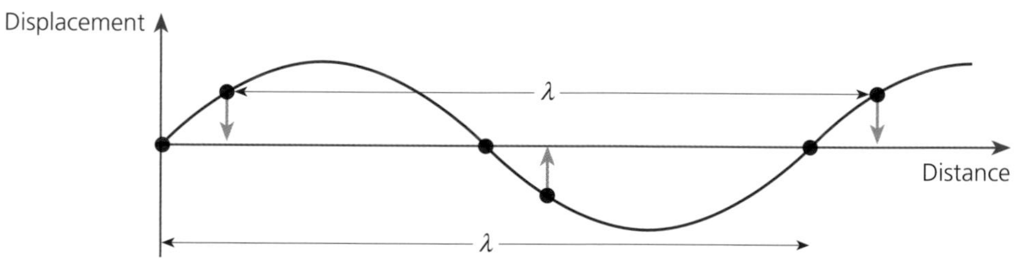

This applies to both transverse and longitudinal harmonic waves.

For the above, we can specify the wavelength λ as the distance between two points on the wave where the particles are in phase.

In a harmonic wave, the displacement of the particles is also harmonic when plotted against time.

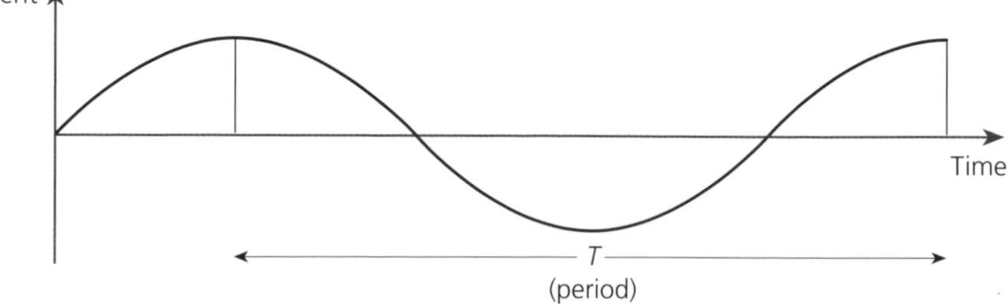

4.3 Progressive harmonic waves

Progressive waves travel from one place to another with a speed $v\,\text{m}\,\text{s}^{-1}$, where the speed is that of some point on the wave — a crest or trough is usually chosen.

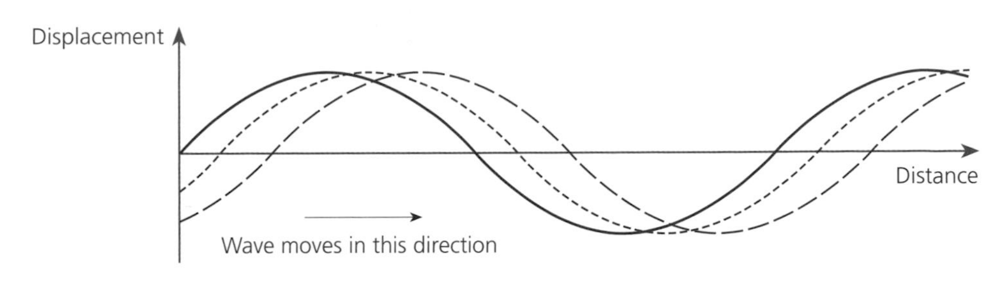

$$\text{speed} = \frac{\text{distance travelled}}{\text{time taken}}$$

Let the distance travelled equal one wavelength of the wave. When considering the oscillation of the particles, the time for this is the time for the particles to make one oscillation. T is the period.

$$\text{speed} = \frac{\text{wavelength}}{\text{period}}, \text{ but frequency} = \frac{1}{\text{period}}$$

$$\text{speed} = \text{frequency} \times \text{wavelength},$$

$$c = f\lambda$$

The equation $c = f\lambda$ can be applied to any shape of travelling wave.

Worked examples

Q1 Sketch a diagram of a square travelling wave, indicating the wavelength. On the same graph, sketch a similar square wave with a phase lag of $\frac{-\pi}{4}$. If the period of the oscillation is 2 s, how long will it take the second wave to reach the current position of the first?

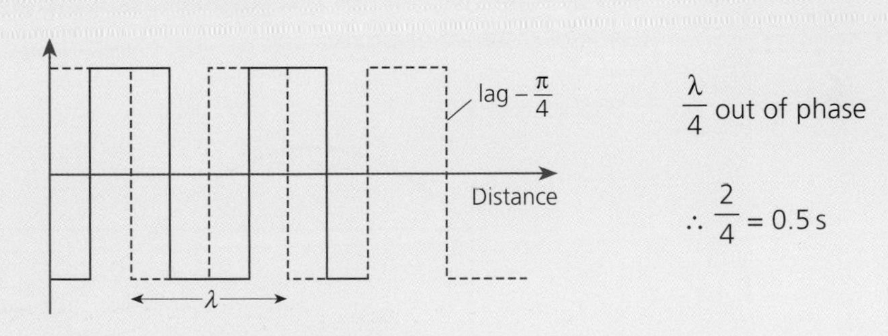

$\frac{\lambda}{4}$ out of phase

$\therefore \frac{2}{4} = 0.5\text{ s}$

Q2 The near infrared part of the electromagnetic spectrum starts at a wavelength of 750 nm and finishes at a wavelength of 2.0 μm. What is the frequency at these two ends?

Take care here with the units of wavelength. See Section 1.

$c = f \times \lambda$ $\quad f = \frac{c}{\lambda} = \frac{3 \times 10^8}{750 \times 10^{-9}} = 4 \times 10^{14}\text{ Hz}$

$f = \frac{c}{\lambda} = \frac{3 \times 10^8}{2 \times 10^{-6}} = 1.5 \times 10^{14}\text{ Hz}$

Q3 The eye is most sensitive to light in the visible part of the spectrum, wavelength 550 nm. What is the frequency and angular frequency of this radiation?

$f = \frac{c}{\lambda} = \frac{3 \times 10^8}{550 \times 10^{-9}} = 5.5 \times 10^{14}\text{ Hz}$

$\omega = 2\pi f = 2 \times \pi \times 5.5 \times 10^{14} = 3.4 \times 10^{15}\text{ rad s}^{-1}$

You should now know:

- **the concept of a travelling wave and the definition of the term 'wavelength'**
- **the difference between longitudinal and transverse waves**
- **how to describe the propagation of harmonic waves**
- **how to obtain the relationship between speed, frequency, and wavelength for a travelling wave**

5 Superposition of waves

In many applications of wave motion, we are interested in what takes place when waves are added together — the **superposition of waves**.

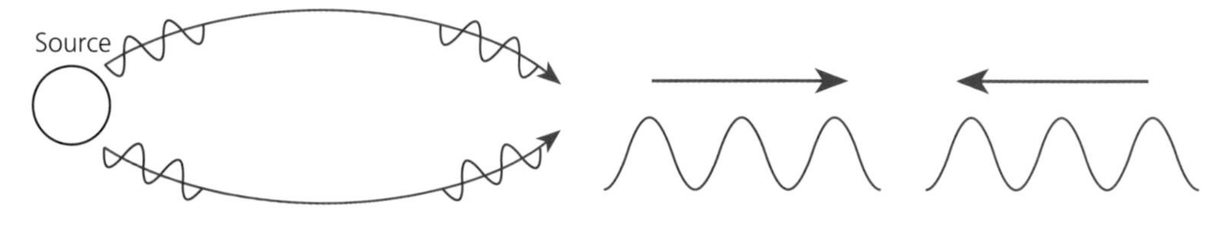

Waves travelling different paths can meet at some point

Waves travelling in opposite directions can meet

The superposition of the waves when they meet is called **interference**.

5.1 Phase and phase difference

We have already seen that it is possible to plot the displacement of harmonic waves in terms of distance or phase.

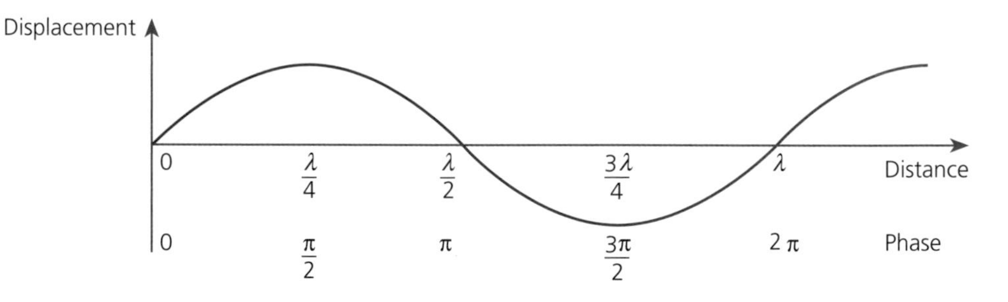

When two waves are superimposed, the relationship between the two waves is usually quoted as a phase difference.

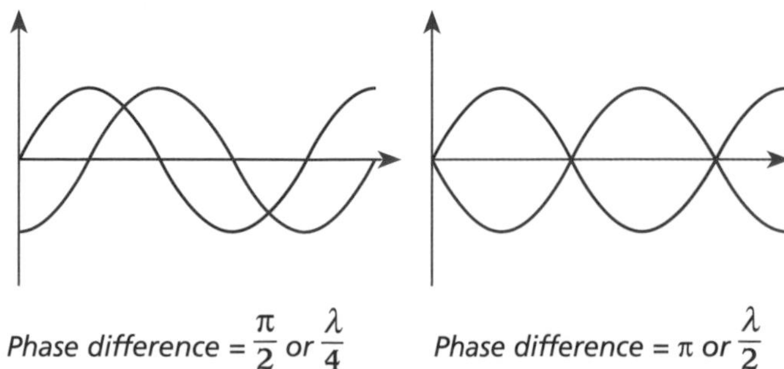

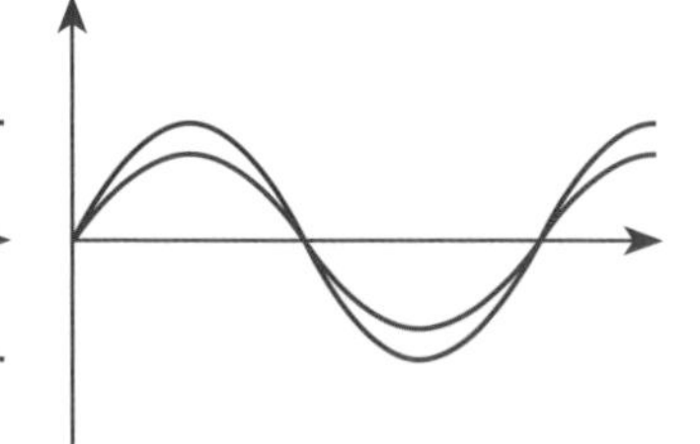

Phase difference = $\frac{\pi}{2}$ or $\frac{\lambda}{4}$

Phase difference = π or $\frac{\lambda}{2}$

Phase difference = 2π or λ – the waves are in phase

5.2 Coherent waves

In many of the applications of superposition, it is important that the waves leaving a source have a fixed phase relationship with one another.

If the waves leaving two sources have zero or a constant phase difference, the sources are said to be coherent.

5.3 Path difference and phase difference

If two waves leave a coherent source, travel different paths and then meet at some point, they will have a **phase difference** depending on the difference in the paths they have travelled — the **path difference**.

We can take a snapshot of the waves at any time and relate distance along the paths, the path length, to the wavelength or the phase. If the paths are different, the waves will arrive at some point with a phase difference related to the path difference.

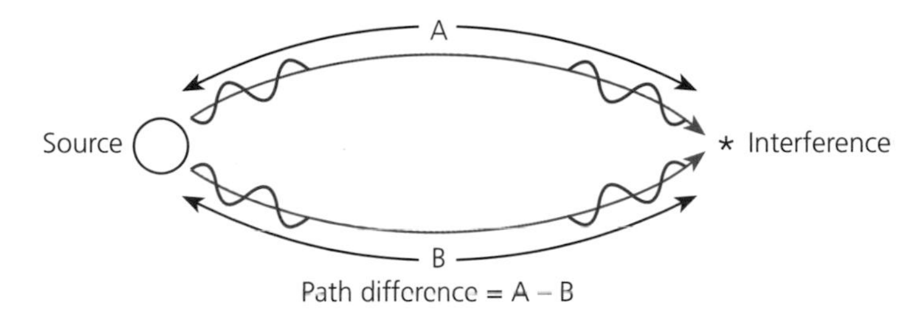

$$\textbf{phase difference} = \frac{2\pi}{\lambda} \times \textbf{path difference}$$

Constructive interference

This occurs when the phase difference between the two waves is 0, 2π, 4π, 6π or $2n\pi$, where n = 0, 1, 2, 3, 4 … .

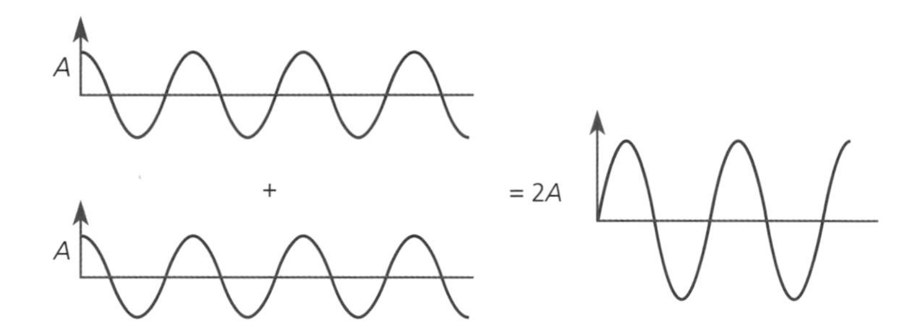

The result is a wave with double the amplitude of the individual waves.

Destructive interference

This occurs when the phase difference between the two waves is π, 3π, 5π or $(2n + 1)\pi$, where n = 0, 1, 2, 3, 4 … .

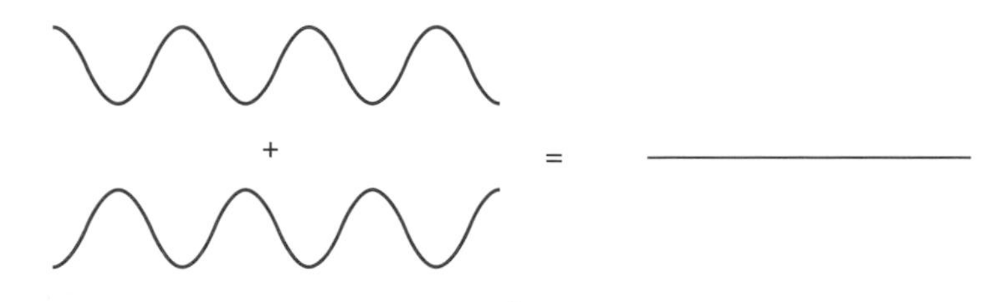

The result is that the two waves cancel each other out.

Worked examples

Q1 The following two square waves are added together. Sketch the form of the resultant wave.

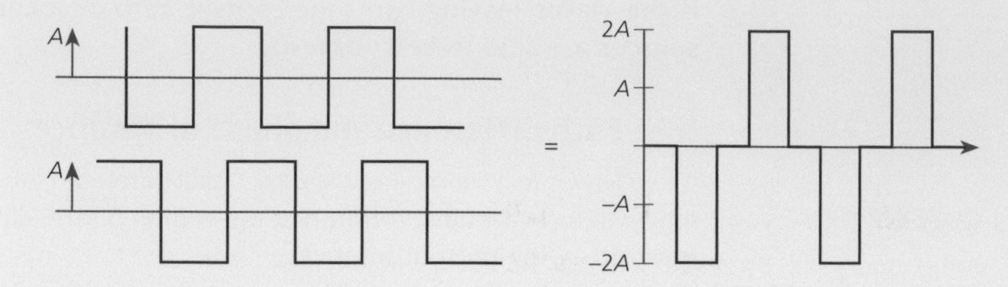

Q2 Green light emitted from a coherent source has a wavelength of 560 nm. If two light beams travel different paths with a path difference of 7560 nm, will they arrive in phase?

Since the calculation uses a ratio of wavelengths, there is no need to convert from nm.

$$\text{phase difference} = \frac{2\pi}{560} \times 7560 = 2 \times \pi \times 13.5$$

$$= 27\pi = (2n + 1)\pi \qquad \therefore \text{ out of phase}$$

Q3 A source emits waves with a wavelength of 5 cm, which travel two different paths. What is the smallest path difference required for the two waves to arrive π out of phase?

$$\pi = \frac{2\pi}{5} \times \text{path difference}$$

$$\text{path difference} = \frac{5 \times \pi}{2 \times \pi} = 2.5\,\text{cm}$$

You should now know:

- **the process of superposition and interference**
- **the definitions of phase, phase difference and coherence**
- **the relationship between path difference and phase difference**

6 *Standing (stationary) waves*

Standing waves are produced when two travelling waves of the same velocity v and wavelength λ moving in opposite directions meet at a point.

The resultant addition of the two waves generates a wave that does not travel — a standing (stationary) wave.

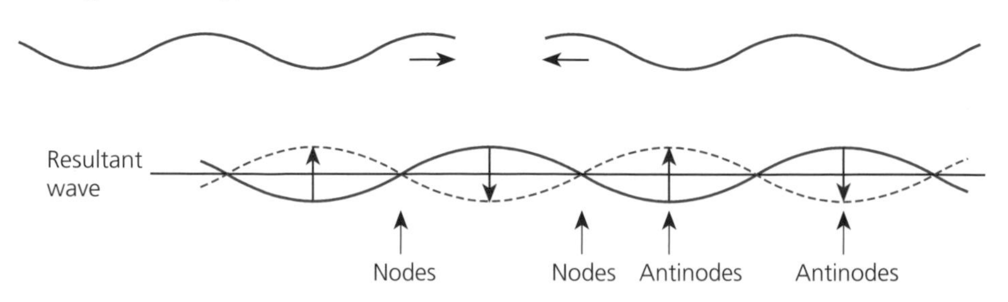

In the resultant wave, the points on the wave where the displacement is always zero are called nodes. Between each node point, where the displacement varies with time, are points called antinodes.

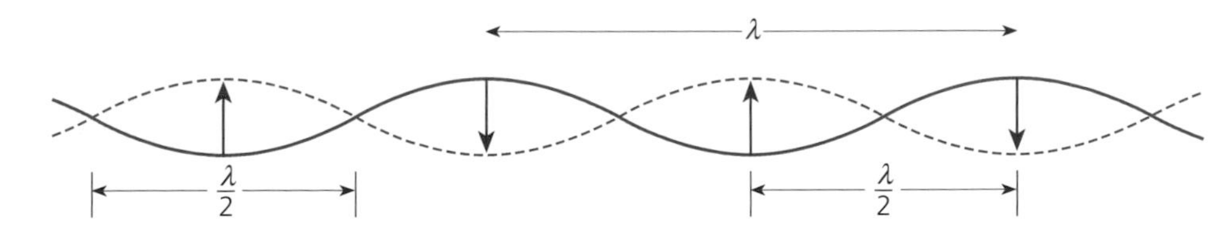

The distance between two adjacent nodes or two adjacent antinodes is equal to $\frac{\lambda}{2}$.

6.1 Practical examples of standing waves

In order to produce a standing wave, we need to generate waves moving in opposite directions. There are several practical examples where this can be achieved.

Stretched string

If a string is stretched between two fixed points, as in a guitar, violin etc., and the string is plucked, waves travelling down the string with velocity v are reflected at the ends and interfere to produce standing waves.

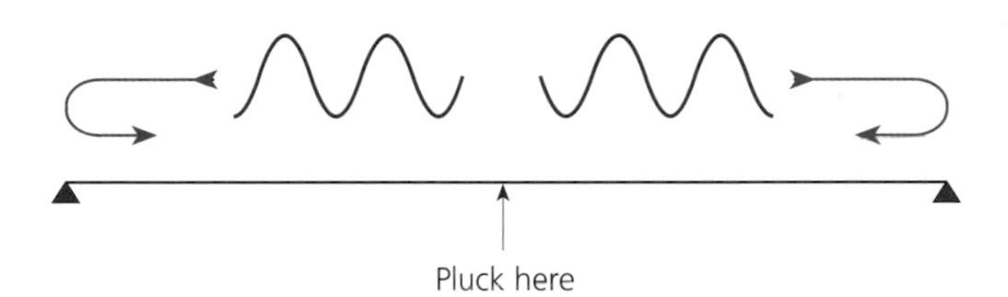

The velocity of the wave down the string is not a constant but depends on the mass per unit length, m, of the string and the tension, F, of the string.

$$\text{speed} = \sqrt{\frac{F}{m}}$$

If a string of length L is fixed at both ends, when a standing wave is produced the ends must be nodes. There are many different standing wave patterns that can be produced with nodes at the ends, and they are given names as follows.

In both stretched strings and organ pipes, the fundamental gives the pitch of the note.

Fundamental or 1st harmonic

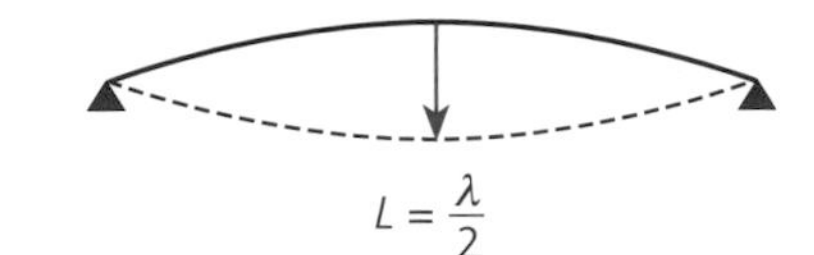

$$\text{frequency} = \frac{v}{2L} = f_0$$

1st overtone or 2nd harmonic

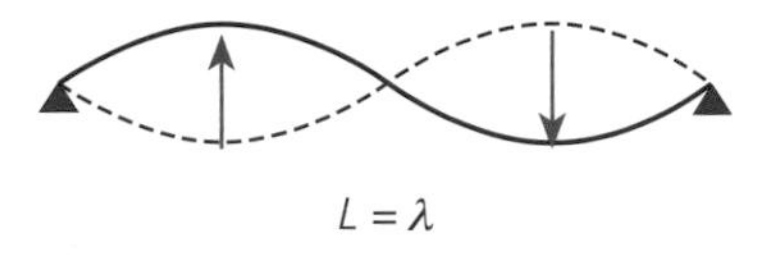

$$\text{frequency} = \frac{v}{L} = 2f_0$$

2nd overtone or 3rd harmonic

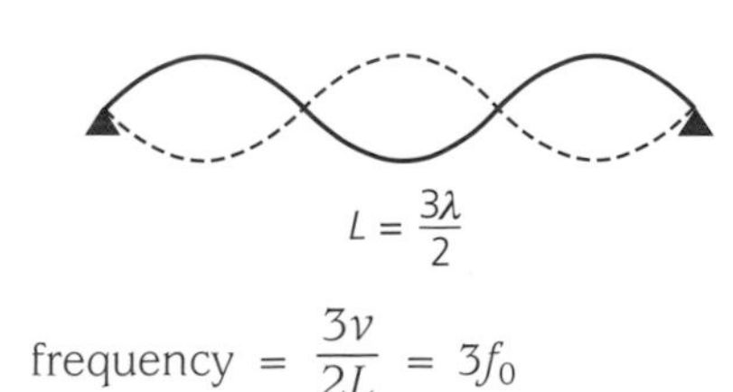

$$\text{frequency} = \frac{3v}{2L} = 3f_0$$

3rd overtone or 4th harmonic

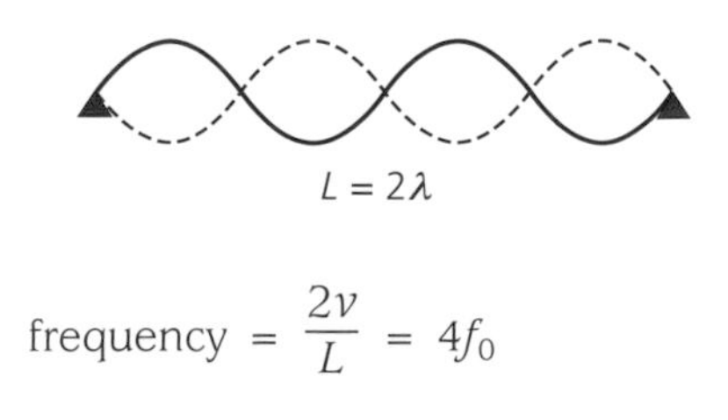

$$\text{frequency} = \frac{2v}{L} = 4f_0$$

When a string is plucked, some harmonics are more dominant than others.

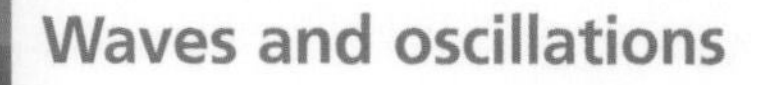

Standing waves in pipes

Sound waves can generate standing waves in pipes. In the case of pipes, we must consider two different pipe constructions: open and closed pipes.

Even when the pipe is open, the sound wave is partially reflected at the open end to generate the standing wave.

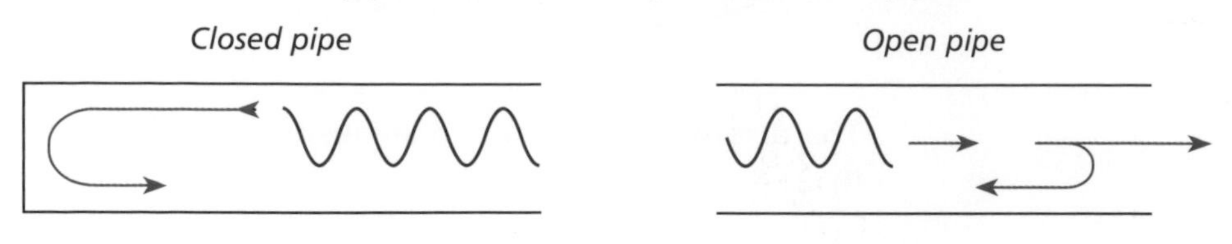

In the closed pipe, the closed end must always be a node and the open end an antinode, while in the open pipe, both ends are antinodes.

Remember, sound is a longitudinal wave motion.

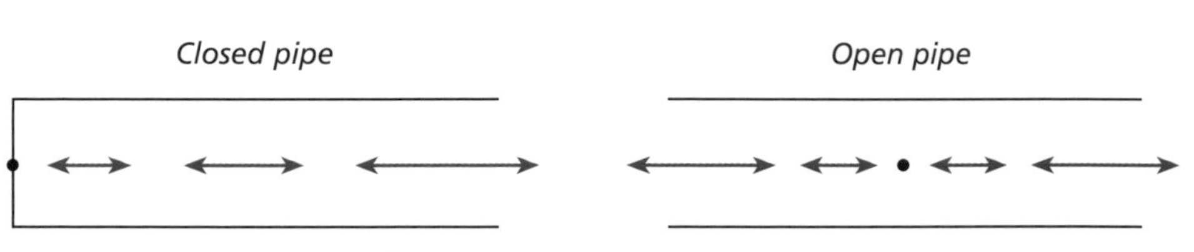

The above are often drawn showing the displacement of the air particles.

In real pipes, the pipe appears longer than its true length and a small end correction must be applied.

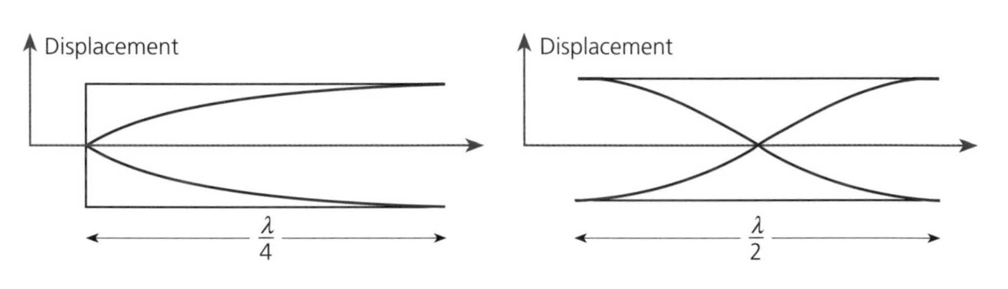

The above diagrams show the fundamental or first harmonic. Higher harmonics are given by a similar process to that in the string, remembering to place nodes or antinodes at the ends of the pipe depending on whether the pipe is open or closed.

Musical instruments all sound different when playing the same note because of the different relative magnitudes of the harmonics produced. The fundamental gives the pitch of the note.

Worked examples

Q1 The G string on a violin is tuned to a frequency of 196 Hz. What are the frequencies of the second and fourth overtones?

$v = f\lambda$ $\qquad$ $v = 196 \times 2L$

$f = \frac{3v}{2L} = 3 \times 196 = 588\text{ Hz}$ $\qquad$ $f = \frac{5v}{2L} = 5 \times 196 = 980\text{ Hz}$

Q2 A string of length 0.6 m and mass per unit length 0.001 kg m^{-1} vibrates at a frequency of 60 Hz when vibrating at the 2nd harmonic. Calculate the tension in the string.

2nd harmonic has frequency $= \frac{v}{L} = 60 \qquad = \frac{v}{0.6}$

$\therefore v = 36\text{ m s}^{-1}$

$v = \sqrt{\frac{F}{m}} \qquad 36 = \sqrt{\frac{F}{m}} \qquad F = 36^2 \times m = 36^2 \times 0.001 = 1.3\text{ N}$

Q3 Sound waves have a velocity of 330 m s^{-1}. What must be the lengths of closed organ pipes to generate notes of frequency matching the top two strings on a violin, which are tuned to pitches of frequency 659 Hz and 440 Hz?

(i) Using $v = f\lambda$: $330 = 659 \times \lambda$ $\lambda = 0.5$ m

length of pipe $= \frac{0.5}{4} = 0.125$ m

(ii) Using $v = f\lambda$: $330 = 440 \times \lambda$ $\lambda = 0.75$ m

length of pipe $= \frac{0.75}{4} = 0.188$ m

You should now know:

- **how standing waves are produced and the difference between nodes and antinodes**
- **the form of standing waves in strings, and in closed and open pipes**
- **the equation for the velocity of a wave in a stretched string**
- **the relationship between fundamentals and overtones**

7 *Reflection and refraction*

This section deals with what happens when a wave meets a boundary between two different media. The rules are applicable to any wave motion but find their greatest application when dealing with light waves. When light waves strike a boundary between two media, they can be **reflected**, **transmitted** into the medium and **refracted**, or **absorbed**.

When considering reflection and refraction, we need to introduce a parameter called the **normal** to the surface.

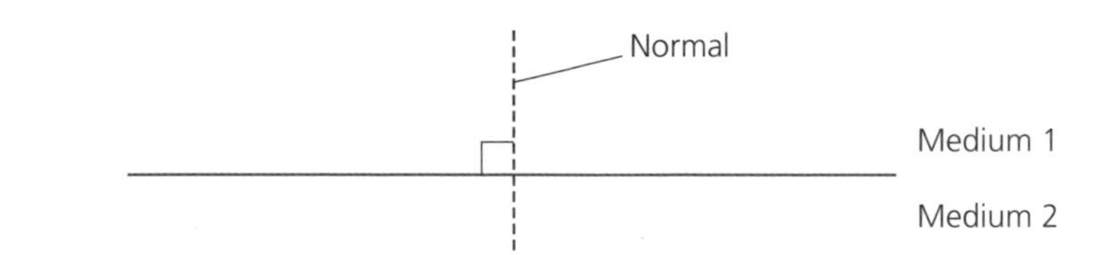

7.1 Reflection

Consider the case of reflection from a plane surface such as a mirror.

The incident ray, the reflected ray and the normal, all lie in the same plane.

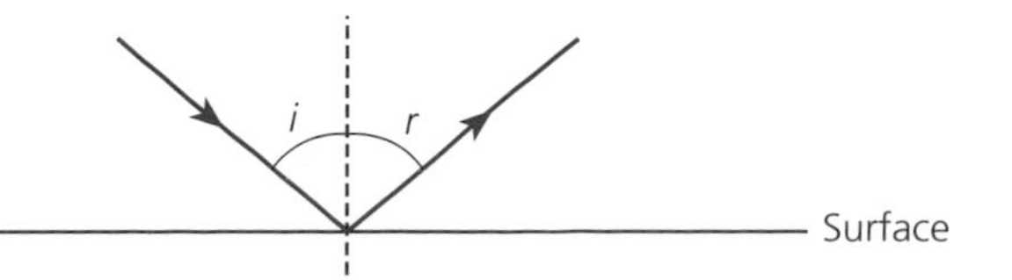

The angle between the incident ray and the normal is equal to the angle between the reflected ray and the normal.

Formation of an image

When an image is formed, the image appears to be behind the mirror. This is called a virtual image.

A virtual image does not exist. It cannot be focused on a sheet of paper. The image is inverted.

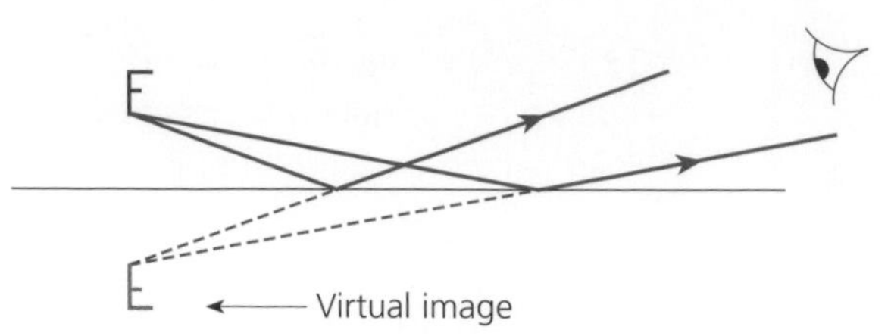

This diagram is reversible. Rays travelling in the reverse direction are refracted away from the normal. Note that optical density refers to the speed of light in the medium: greater optical density means slower speed.

7.2 Refraction

Refraction is the process that takes place when light travels from one medium to another with a different optical density and its path bends. The amount the light bends is dependent on the velocity of light in the two media.

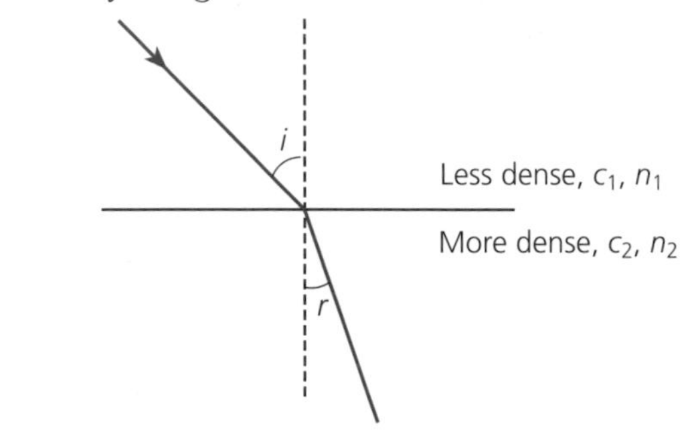

$$\frac{\text{sine of the angle of incidence}}{\text{sine of the angle of refraction}} = \frac{\text{speed of light in the less dense medium } (c_1)}{\text{speed of light in the more dense medium } (c_2)}$$

$$= \frac{n_1}{n_2}$$

- n_1 = the refractive index of the less dense medium
- n_2 = the refractive index of the more dense medium

In many cases, we are considering light striking the boundary between air and, say glass, which to within very small limits is the same as that between vacuum and glass. In this case, we just talk about n, the refractive index of the glass, since $n_1 = 1$ for a vacuum and 1.0003 for air.

For air or a vacuum, c_1 = the velocity of light in vacuum = $c = 3 \times 10^8$ m s^{-1}.

Snell's law

$$\frac{\sin i}{\sin r} = \frac{c}{c_2} = \frac{n_2}{n_1} = n_2 \text{ for air or a vacuum as the less dense medium}$$

The refractive index of a medium also depends on the wavelength of light. In a prism, different wavelengths are refracted through different angles.

Refraction in action

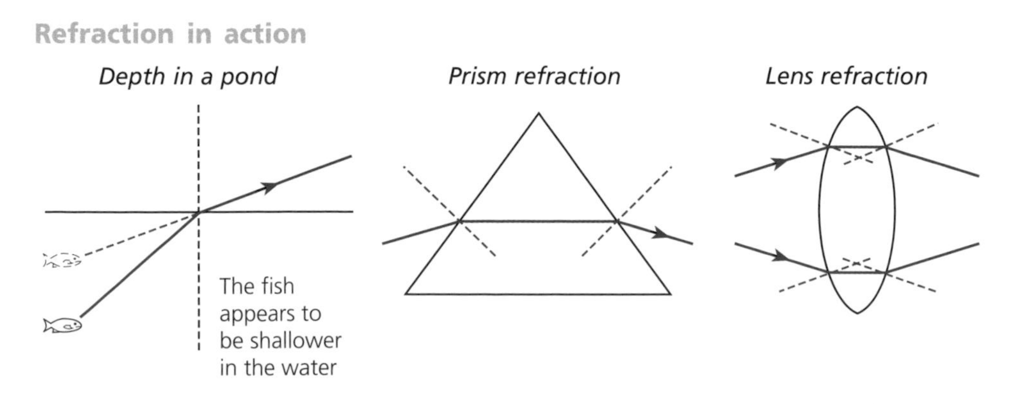

7.3 Total internal reflection

When light travels from an optically dense to a less optically dense medium, refraction away from the normal takes place. As the angle of incidence increases,

so the angle of refraction increases, until it eventually reaches 90° and the light ray grazes the surface. A further slight increase causes the ray to be totally internally reflected.

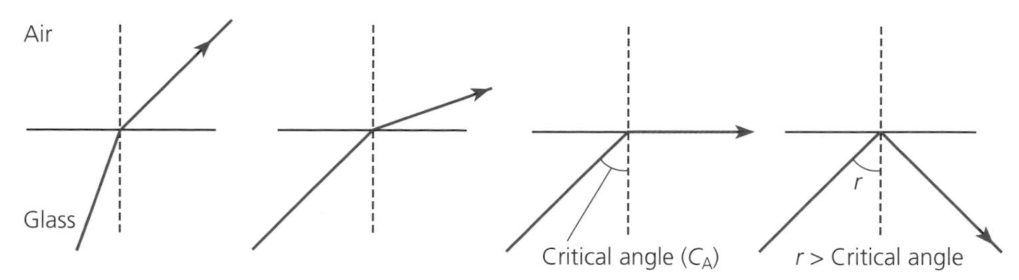

When light leaving grazes the surface, the incident angle is called the critical angle.

$$\text{refractive index} = \frac{\sin i}{\sin r} = n_2 \qquad i = 90° \text{ so } \sin i = 1 \qquad r = C_A \qquad \frac{1}{\sin C_A} = n_2$$

Total internal reflection is used to keep a light signal in an optical fibre.

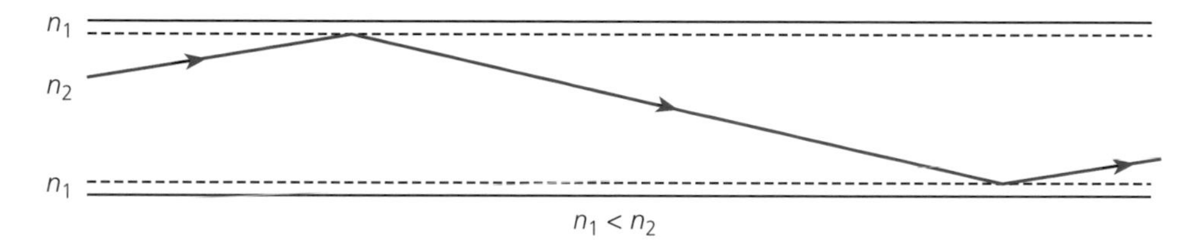

Worked examples

Q1 Diamond has a refractive index of 2.417. Light is incident on an air–diamond boundary at an angle of 35°. What is the angle of refraction and the velocity of light in diamond?

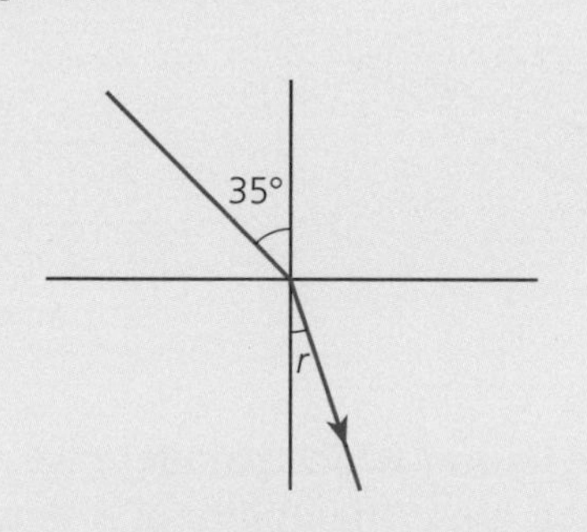

$$\frac{\sin i}{\sin r} = n \qquad \frac{\sin 35}{\sin r} = 2.417$$

$$\sin r = \frac{\sin 35}{2.417} \qquad r = 13.7°$$

Velocity of light in diamond:

$$2.417 = \frac{c}{c_d} = \frac{3 \times 10^8}{c_d}$$

$$c_d = 1.2 \times 10^8\ \text{m s}^{-1}$$

Q2 Light passes symmetrically through a glass prism with a vertex angle of 40°. The refractive index of the glass is 1.52. What is the total deviation of the light passing through the prism?

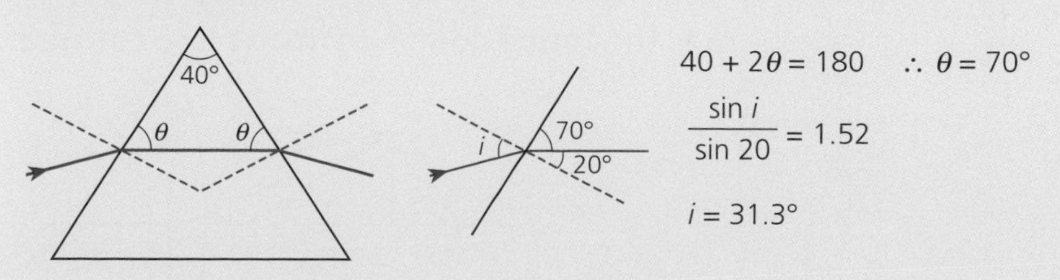

$$40 + 2\theta = 180 \qquad \therefore\ \theta = 70°$$

$$\frac{\sin i}{\sin 20} = 1.52$$

$$i = 31.3°$$

deviation at the first surface = 31.3 – 20 = 11.3°

total deviation = 2 × 11.3 = 22.6°

Q3 Optical fibres use total internal reflection and a critical angle of 85°. The outer fibre has a refractive index of 1.520. What must be the refractive index of the inner core?

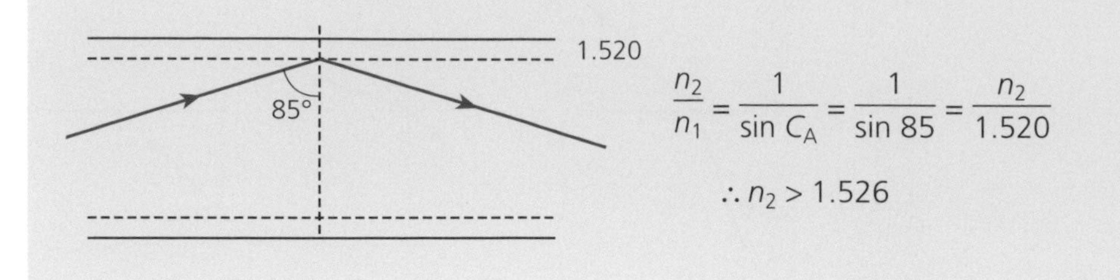

$$\frac{n_2}{n_1} = \frac{1}{\sin C_A} = \frac{1}{\sin 85} = \frac{n_2}{1.520}$$

$$\therefore n_2 > 1.526$$

You should now know:

- **the difference between reflection and refraction**
- **the normal to a surface, angle of incidence, reflection and refraction, Snell's law**
- **total internal reflection and the definition of critical angle**

8 *Lens types*

In the previous section, we saw that refraction takes place in a lens. Provided the lens surfaces are close together and do not have small radius curves, we can apply a simple theory, called thin lens theory, to the passage of light through the lens.

Converging lens (convex lens)

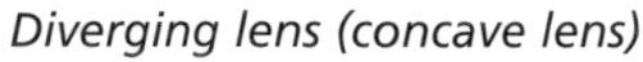

Diverging lens (concave lens)

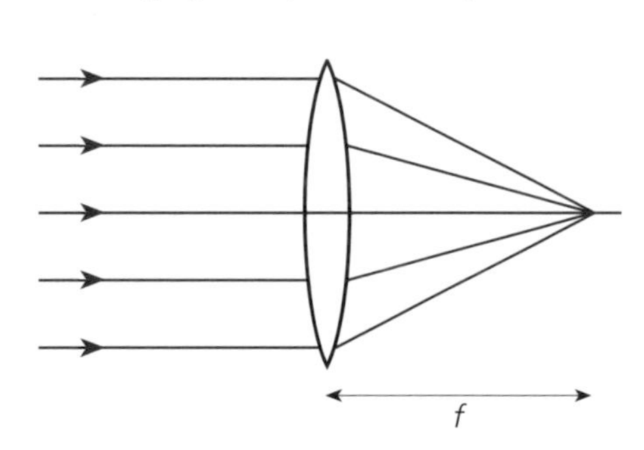

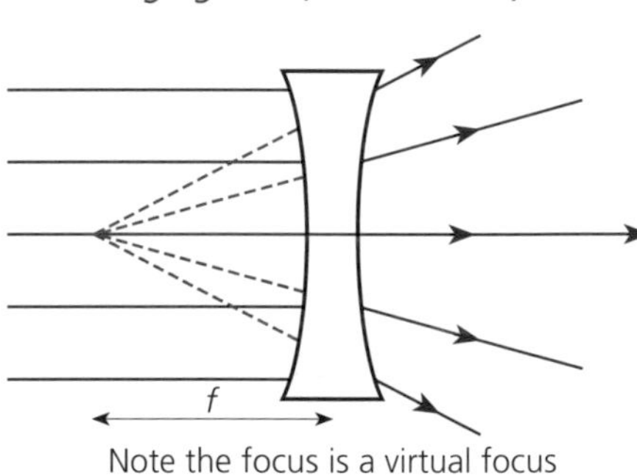

Practical lens systems are more complicated than this and are often constructed from many lenses joined together using glass of different refractive index.

Power of a lens

The distance from the lens at which parallel rays are brought to a focus is called the focal length of the lens. The ability of a lens to focus parallel rays a short distance from the lens is called the power of the lens.

$$\text{power of the lens (units, dioptre)} = \frac{1}{\text{focal length of the lens}}$$

When combining two lenses of focal length f_1 and f_2, the total length of the combination is:

$$\frac{1}{f} = \frac{1}{f_1} + \frac{1}{f_2}$$

8.1 The formation of an image by a converging lens

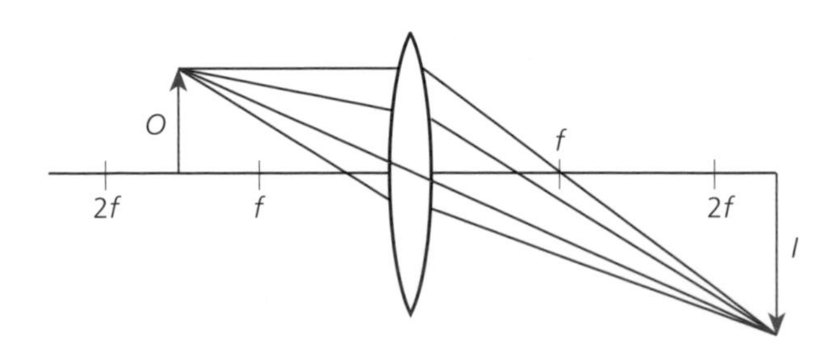

Note the following:

- a ray through the centre of the lens is not deviated
- a ray parallel to the axis is refracted to pass through the focus

We can use these rules to construct ray diagrams.

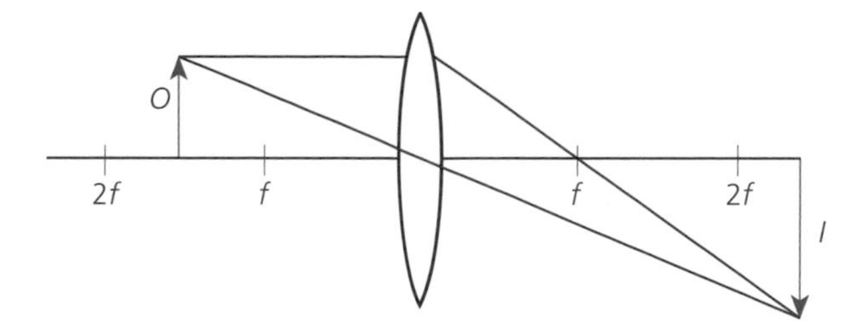

In this example the object is between f and $2f$.

8.2 Magnification

Using triangles similar to those in the above simple ray diagram:

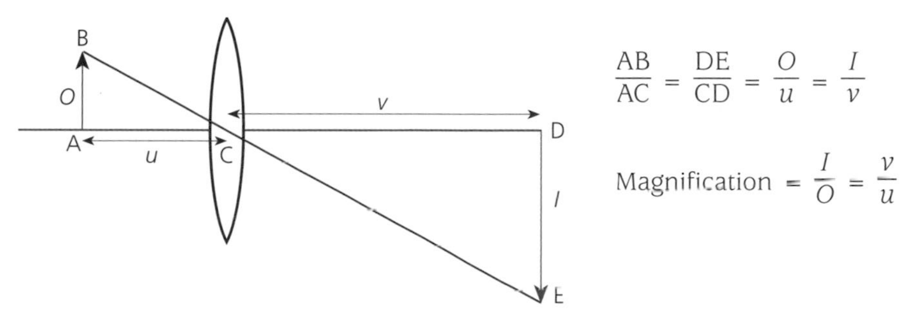

$$\frac{\text{AB}}{\text{AC}} = \frac{\text{DE}}{\text{CD}} = \frac{O}{u} = \frac{I}{v}$$

$$\text{Magnification} = \frac{I}{O} = \frac{v}{u}$$

8.3 Lens formula

Use of the lens formula requires a convention to deal with virtual images, and we shall use the most common convention: **real is positive.**

The relationship between v and u can also be calculated mathematically using the thin lens formula.

$$\frac{1}{f} = \frac{1}{v} + \frac{1}{u}$$

f = focal length of the lens
u = object distance
v = image distance

Image configurations

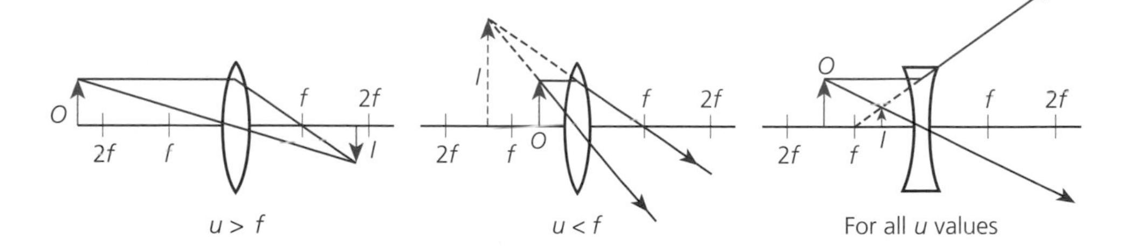

8.4 Eyesight correction

Note the refraction at the cornea as well as the lens.

The lens of the eye changes shape to bring into focus objects at different distances.

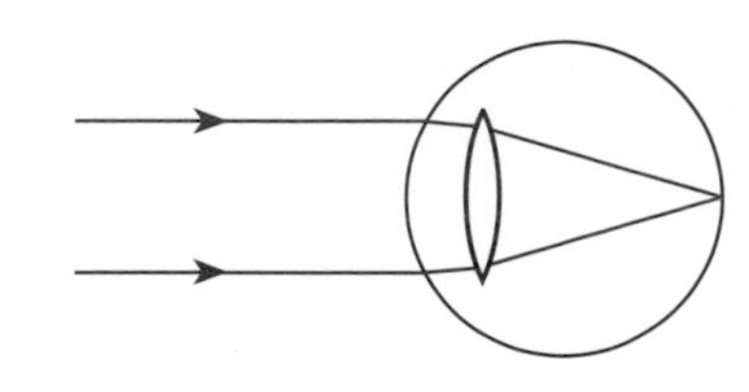

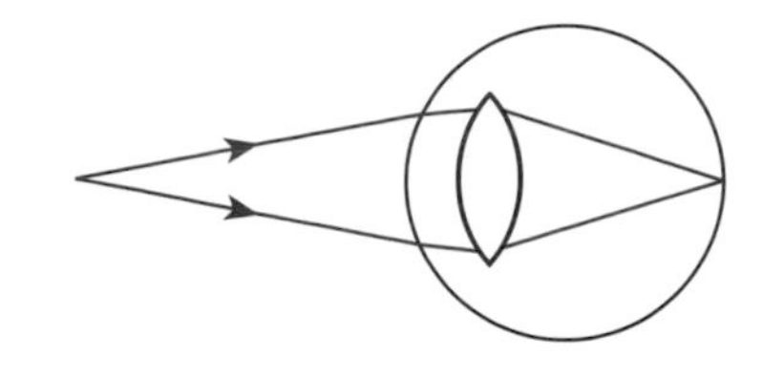

Short-sighted people see objects close to the eye clearly. Long-sighted people see distant objects clearly. Long sight is one of the most common eye defects.

Eye defects and correction

Short sight – unable to focus distant objects

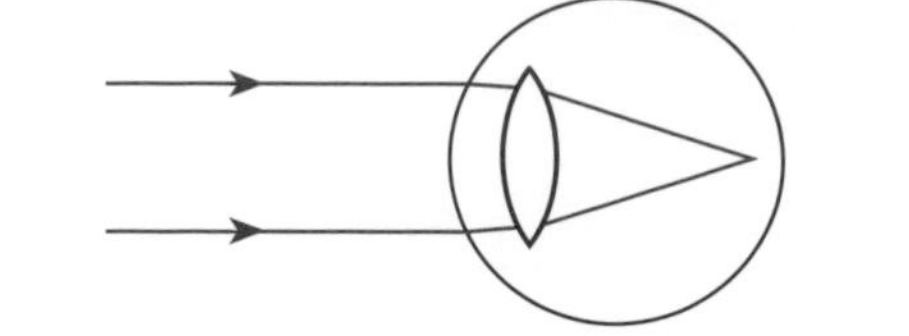

Long sight – unable to focus close objects

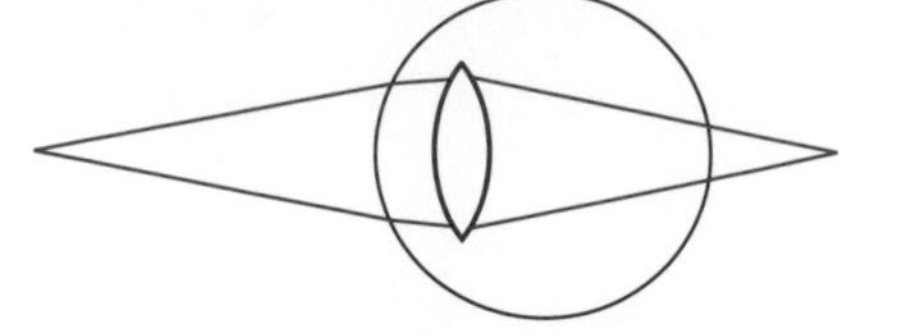

Correction method for short sight

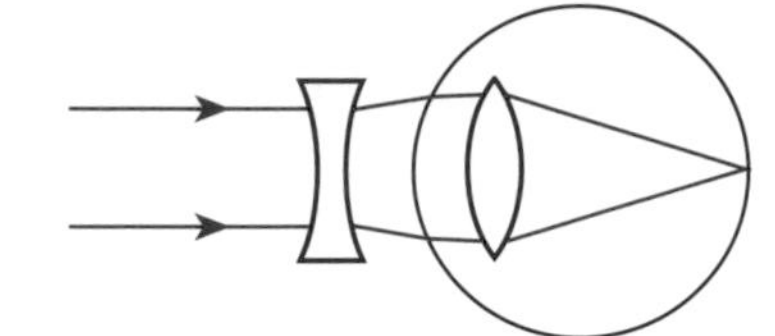

Correction method for long sight

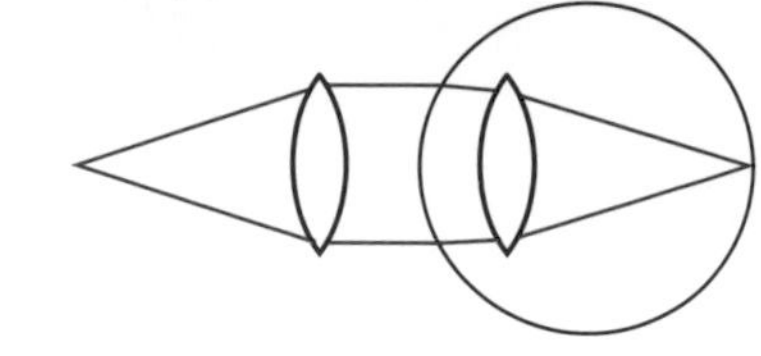

Worked examples

Q1 A projector with a lens of focal length 20 cm is used to focus an object 1.5 cm high on a screen. The object is placed 25 cm from the lens. At what position should the screen be placed and what is the magnification produced?

We can work in cm in this question, but remember that the answer will be in cm.

$$\frac{1}{f} = \frac{1}{v} + \frac{1}{u} \qquad \frac{1}{20} = \frac{1}{25} + \frac{1}{v}$$

$$\frac{1}{v} = \frac{1}{20} - \frac{1}{25} = \frac{1}{100} \qquad v = 100\text{ cm}$$

$$\text{magnification} = \frac{v}{u} = \frac{100}{25} = 4 \qquad \text{object height} = 6\text{ cm}$$

Q2 Virtual magnified images are seen clearly between ∞ and 160 mm through a thin converging lens with a focal length of 80 mm. What is the range of positions of the object?

$$\frac{1}{f} = \frac{1}{v} + \frac{1}{u} \qquad \frac{1}{80} = -\frac{1}{\infty} + \frac{1}{u} \qquad \therefore u = 80\text{ mm}$$

$$\frac{1}{f} = \frac{1}{v} + \frac{1}{u} \qquad \frac{1}{80} = \frac{1}{160} + \frac{1}{u} \qquad \frac{1}{u} = \frac{3}{160}$$

$$u = 53.3\text{ mm}$$

$$\therefore 53.3\text{ mm} \le u \le 80\text{ mm}$$

Q3 A long-sighted man can just see clearly an object placed a distance of 1 m from his eye. Closer objects are out of focus. The diameter of his eye ball is 2 cm. What focal length lens must be used to allow him to see objects placed a distance 20 cm from his eye?

Calculate the shortest focal length of the eye lens:

$$\frac{1}{f_2} = \frac{1}{2} + \frac{1}{100} = \frac{51}{100} \qquad f_2 = \frac{100}{51}\text{ cm}$$

Calculate the focal length required to see the object:

$$\frac{1}{f} = \frac{1}{2} + \frac{1}{20} = \frac{11}{20} \qquad f = \frac{20}{11}\text{ cm}$$

$$\frac{1}{f} = \frac{1}{f_1} + \frac{1}{f_2} \qquad \frac{11}{20} = \frac{1}{f_1} + \frac{51}{100} \qquad \frac{1}{f_1} = \frac{11}{20} - \frac{51}{100} = \frac{4}{100}$$

$$f_1 = 25\text{ cm}$$

You should now know:
- **the difference between converging, diverging, focal length and power of a lens**
- **how to construct simple ray diagrams**
- **magnification and the use of the lens formula**
- **short and long sight and its correction**

9 Diffraction and interference

In many practical situations, diffraction is first required to send waves along different paths so that, when they recombine, interference takes place.

The words 'diffraction' and 'interference' are often used incorrectly, and it is important to distinguish between the two.

Diffraction is the bending of waves around an object.

Interference is the superposition of waves with particular phase differences, producing constructive and destructive interference.

9.1 Practical observation of diffraction

The observation of diffraction effects depends on the size of the object relative to the wavelength of the interacting waves. Consider diffraction at a single slit:

These diagrams could represent water waves entering a harbour or light passing through a narrow slit.

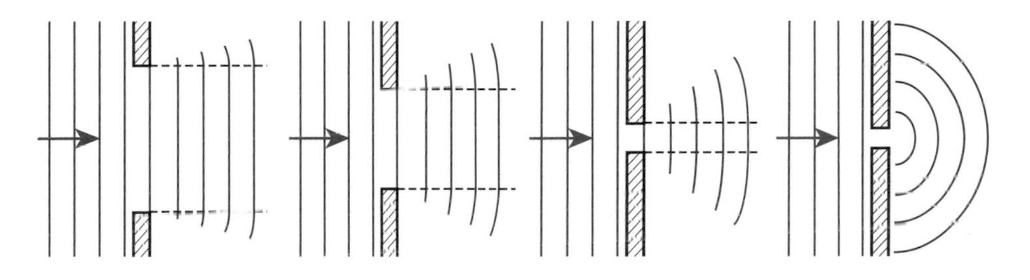

The 'same order' means to within approximately the same power of 10.

As a simple rule, diffraction effects are observed when the size of the object, the aperture in this case, is of the same order as the wavelength of the waves.

9.2 Diffraction by other objects

Visible light has wavelengths of 400–700 nm, so objects for diffracting light must have a similar size to these values.

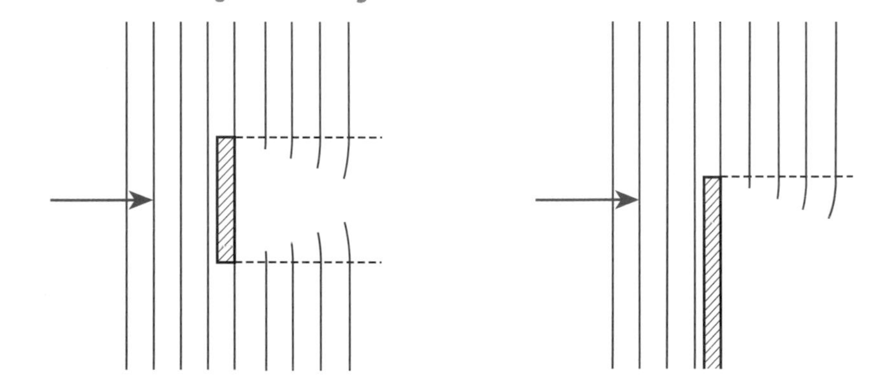

9.3 Single-slit optical diffraction

To observe single-slit diffraction:
- the slit must have a size of approximately 10^{-5} m
- the source must be monochromatic, i.e. have a single wavelength
- the source must be coherent, i.e. all the waves from the source must have a constant phase

See Section 5.2 for a definition of coherent.

Coherent sources can be obtained by restricting the area of a monochromatic source. A laser is a source of monochromatic, coherent radiation.

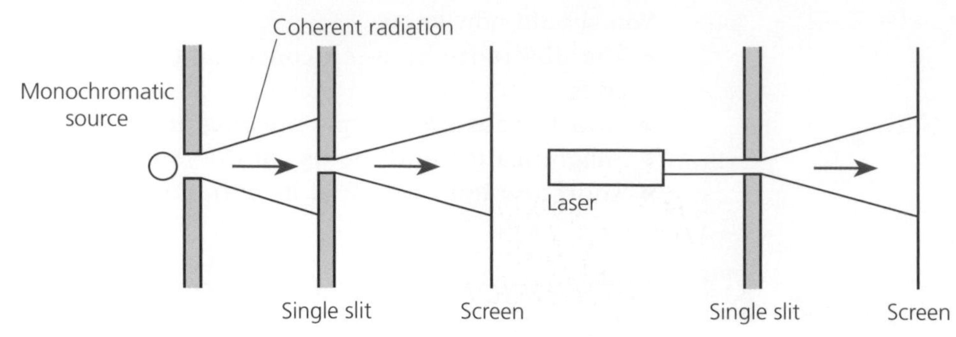

The intensity pattern that is produced on the screen by the interference of waves diffracted from different parts of the slit looks like this:

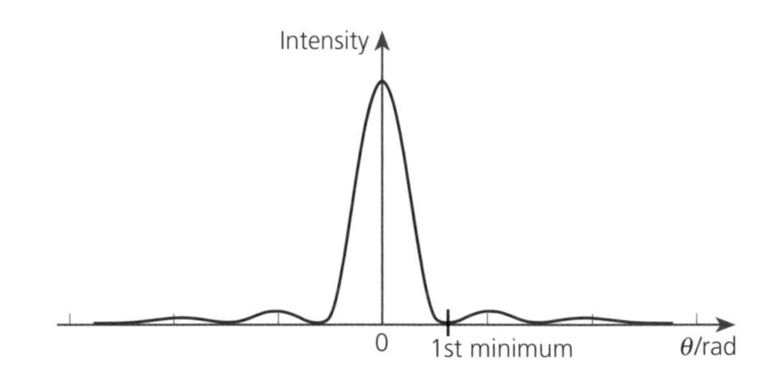

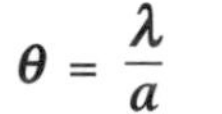

You would not be expected to prove this, but be aware of the general form of the equation.

In single-slit diffraction, the angle in radians of the first minimum on the screen is:

$$\boldsymbol{\theta = \frac{\lambda}{a}} \qquad \boldsymbol{a} = \textbf{slit width} \qquad \boldsymbol{\lambda} = \textbf{wavelength of the light}$$

Diffraction by a circular aperture

With a circular aperture, the diffraction pattern is a set of concentric circles, and the intensity pattern across the diameter of these circles is similar to that of the single slit.

The intensity pattern is slightly larger, with the first minimum now at an angle.

$$\theta = \frac{1.22\lambda}{a}$$

a = the diameter of the circular aperture

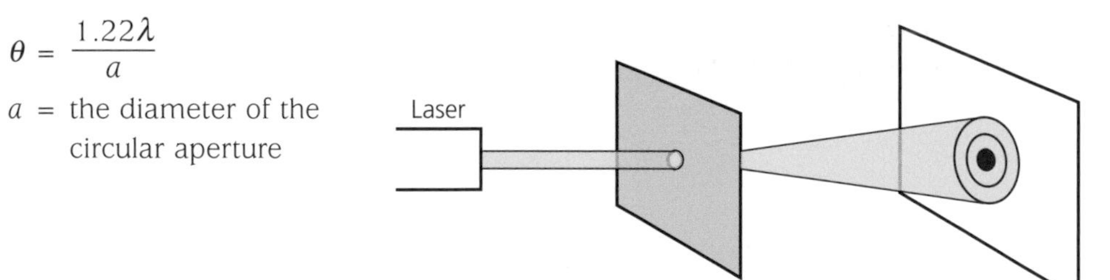

Telescopes looking at distant stars produce images that are circular diffraction patterns produced by the lens aperture. Two stars close together produce overlapping diffraction patterns. At a certain separation of the stars, the two diffraction patterns overlap so much that the two images appear as one image — the two stars appear as one star.

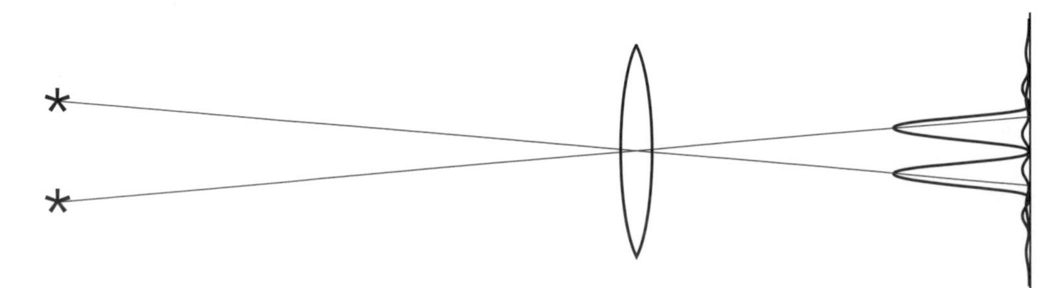

Worked examples

Q1 Light from a laser has a wavelength of 632 nm and is incident normally on a slit of width 0.05 mm. Calculate the angle of the first minimum in the diffraction pattern. What is the distance between the two minima either side of the central maximum on a screen 3 m away?

$$\theta = \frac{\lambda}{a} = \frac{632 \times 10^{-9}}{0.05 \times 10^{-3}} = 1.26 \times 10^{-2} \text{ rad}$$

Use $x = r2\theta$:

$$x = 3 \times 2 \times 1.26 \times 10^{-2}$$

$$= 7.6 \times 10^{-2} \text{ m}$$

Q2 Sound waves of frequency 550 Hz pass normally through a door of width 0.8 m. Calculate the angle of the first minimum on the other side of the door.

Use $v = f\lambda$: $331 = 550 \times \lambda$ $\quad \therefore \lambda = \frac{331}{550} = 0.60 \text{ m}$

$$\theta = \frac{\lambda}{a} = \frac{0.60}{0.80} = 0.75 \text{ rad}$$

Q3 Water waves of wavelength 30 m meet a harbour wall normally, and the entrance has a width of 45 m. What is the angular width of the central beam in degrees?

$$20 = \frac{2\lambda}{a} = \frac{2 \times 30}{45} = 1.3 \text{ rad}$$

$$\text{In degrees} = \frac{1.3 \times 360}{2 \times \pi} = 76°$$

Angular width is the angle between the first minima either side of the central maximum.

You should now know:

- **the difference between interference and diffraction**
- **how a single-slit diffraction pattern is produced, and the angle of the first minimum**
- **the diffraction pattern due to a circular aperture and the angle of the first minimum**
- **how diffraction affects the use of telescopes**

10 Double slits and gratings

10.1 Young's double-slit interference pattern

Two slits of width a separated by a distance d are illuminated with monochromatic coherent radiation, d being greater than a. The following Young's double-slit pattern is obtained.

You should compare the profile of the interference pattern with a single-slit pattern. They are the same, with the first minimum at an angle $\theta = \frac{\lambda}{a}$, where a is the width of one of the slits.

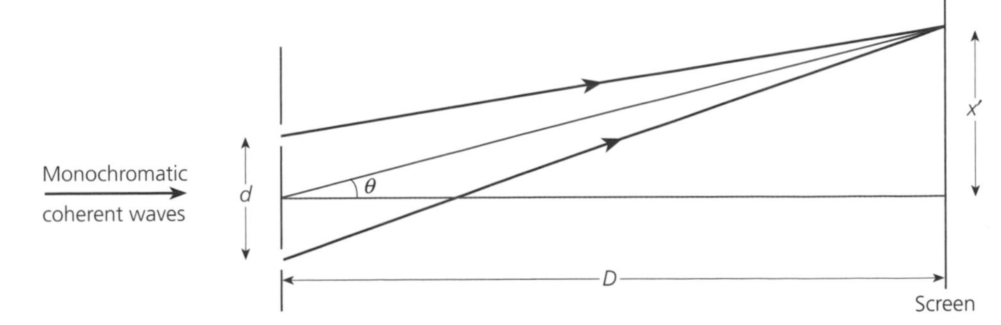

The pattern on the screen is as follows:

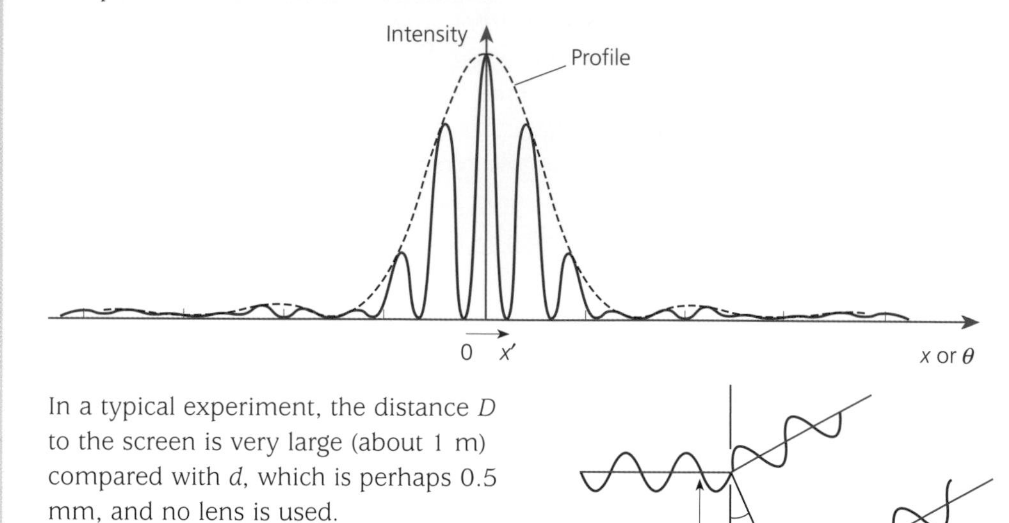

In this calculation θ is very small, in which case sin θ = tan θ = θ rad.

In a typical experiment, the distance D to the screen is very large (about 1 m) compared with d, which is perhaps 0.5 mm, and no lens is used.

If the first maximum is at an angle θ, when the extra path travelled by the lower wave is a whole wavelength:

$$\tan\theta = \theta = \frac{x'}{D} \text{ and } \sin\theta = \theta = \frac{\lambda}{d} \text{ hence } \frac{x'}{D} = \frac{\lambda}{d} \qquad x' = \frac{\lambda D}{d}$$

Interference with sound waves

The same experiment can be repeated with sound waves, except that in this experiment the distances across the pattern are larger, owing to the fact that sound waves have a wavelength much greater than that of light.

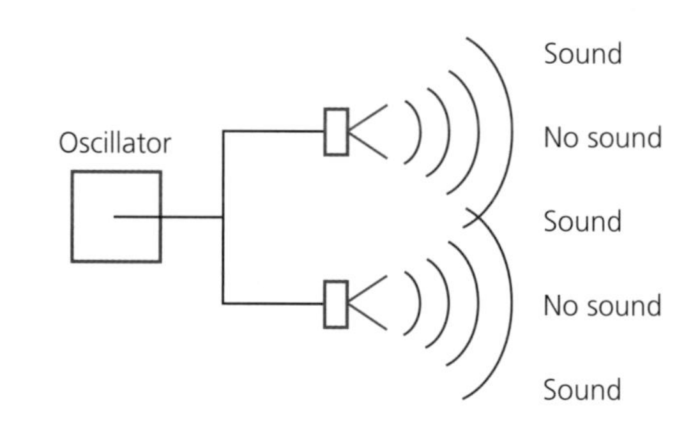

10.2 The diffraction grating

The diffraction grating is a large number of single slits side by side, with slit width a and slit spacing d. d is again greater than a.

In a diffraction grating, a lens is used to bring parallel rays leaving the grating to a focus on the screen.

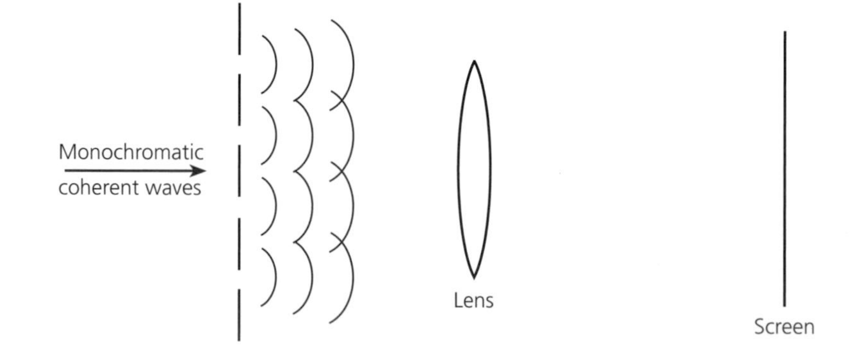

Incident parallel waves are diffracted at each slit. At an infinite distance away from the grating, when they overlap, interference takes place.

The conditions for constructive interference — first-order diffraction

Consider two adjacent slits illuminated with coherent radiation (which is not monochromatic). Light rays leave the slits at some angle θ and will be in phase if the extra path travelled by the lower wave is equal to *one* whole wavelength. When recombined by a lens, constructive interference takes place.

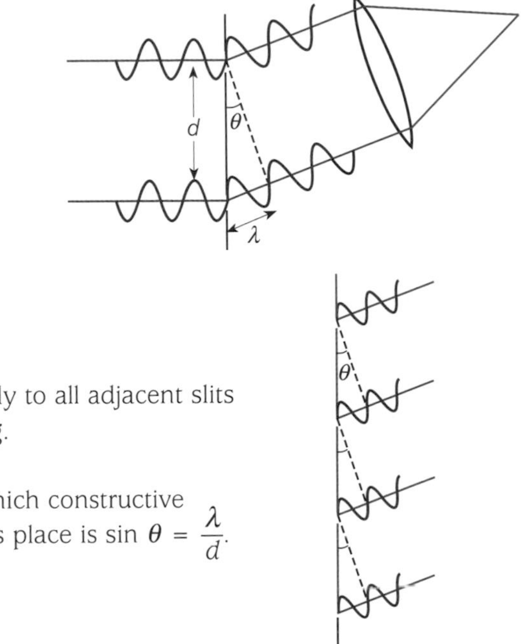

This rule will apply to all adjacent slits across the grating.

The angle θ at which constructive interference takes place is $\sin\theta = \frac{\lambda}{d}$.

Second-order diffraction

If the extra path travelled by the lower wave is equal to *two* whole wavelengths, then, when recombined by a lens, constructive interference again takes place.

The angle θ' at which the above applies is $\sin\theta' = \frac{2\lambda}{d}$.

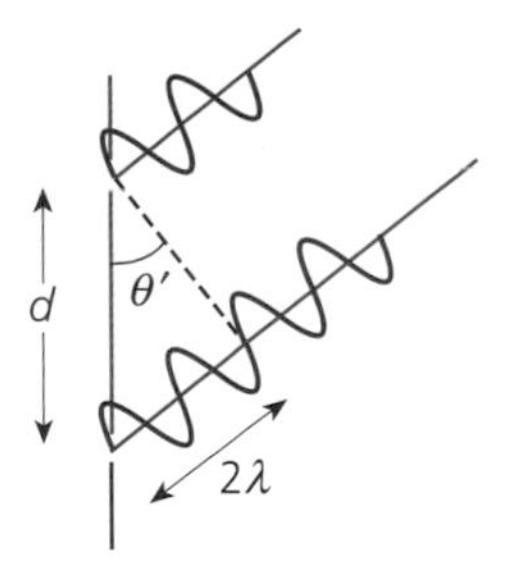

The general rule is that for constructive interference $\sin\theta = \frac{n\lambda}{d}$, where n is an integer 1, 2, 3, 4 … .

If the source contains light of three increasing wavelengths λ_1, λ_2 and λ_3, constructive interference for each takes place at slightly larger angles.

With d constant, in each order $\sin\theta \propto \lambda$.

You should note that the angular spread of the lines increases with the order, making it easier to see the spectrum.

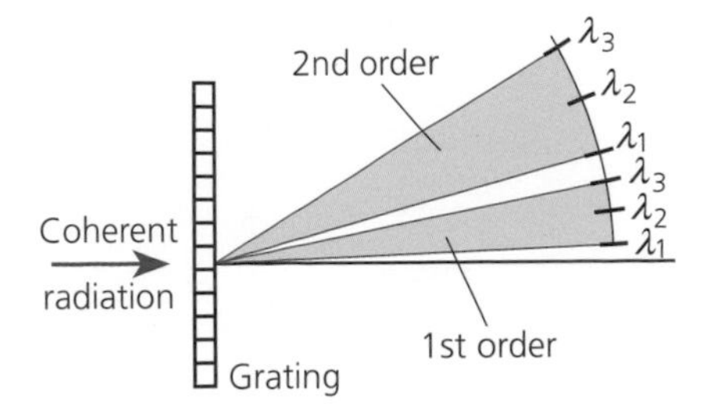

Worked examples

Q1 A monochromatic source of wavelength 550 nm illuminates a double slit of slit spacing $d = 0.5$ mm and slit width $a = 0.05$ mm. What is the spacing of the fringes formed on a screen 1.5 m from the slits? At what angle is the profile of the pattern a minimum?

Use $x = \frac{\lambda D}{d}$: $x = \frac{550 \times 10^{-9} \times 1.5}{0.5 \times 10^{-3}} = 1.7 \times 10^{-3}$ m

Use $\theta = \frac{\lambda}{a}$: $\theta = \frac{550 \times 10^{-9}}{0.05 \times 10^{-3}} = 1.1 \times 10^{-2}$ rad

Q2 A diffraction grating is constructed with 500 lines per mm. What is the angular spread of the visible spectrum, 400 nm to 750 nm in the first order?

500 lines per mm is a spacing of $\frac{1 \times 10^{-3}}{500} = 2 \times 10^{-6}$ m

$\sin\theta = \frac{\lambda}{d} = \frac{400 \times 10^{-9}}{2 \times 10^{-6}} = 0.2$ rad

$\sin\theta = \frac{\lambda}{d} = \frac{750 \times 10^{-9}}{2 \times 10^{-6}} = 0.375$ rad   Angular spread = 0.175 rad

Q3 Using the above grating, would the 750 nm end of the second-order spectrum overlap the 400 nm end of the third-order spectrum?

2nd order $\sin\theta = \frac{2 \times 750 \times 10^{-9}}{2 \times 10^{-6}} = 0.75$ rad

3rd order $\sin\theta = \frac{3 \times 400 \times 10^{-9}}{2 \times 10^{-6}} = 0.60$ rad   Yes, they would overlap.

You should now know:

- **the set-up for showing Young's interference fringes**
- **the equation for the fringe spacing**
- **the diffraction grating and the general equation for constructive interference**

11 *Polarisation*

Transverse waves have the ability to be polarised.

In an **unpolarised wave**, the transverse oscillations lie in any plane.

In a polarised wave, the transverse oscillations lie in one plane.

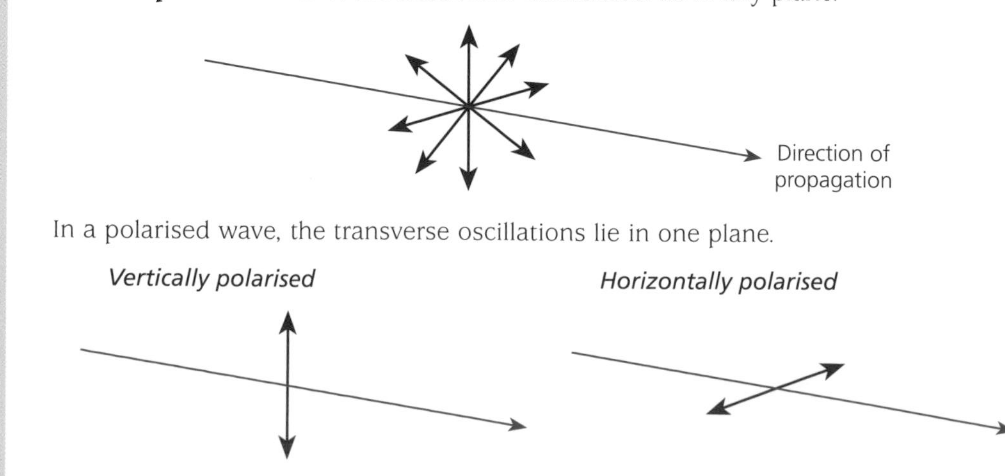

11.1 Polarised light

Light waves from many sources are initially unpolarised. They can be polarised using a polarising filter, of which Polaroid film is one example. Polaroid film is a transparent plastic material consisting of long-chain molecules, which lie in one direction. When unpolarised light passes through the film, light emerging on the other side of the film is polarised.

Simply rotate the Polaroid film through 90° to convert from one form to the other.

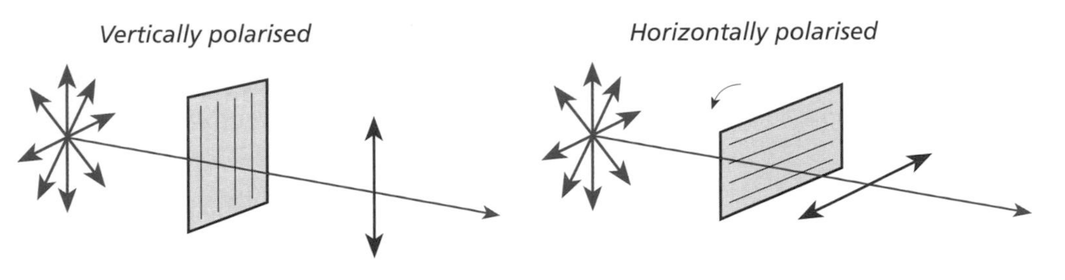

11.2 Polarisation by reflection

Light reflected from a surface is partially polarised in a direction parallel to the surface.

Polaroid sunglasses are designed to reduce reflection by using a Polaroid sheet at a suitable angle to remove the horizontally polarised reflection.

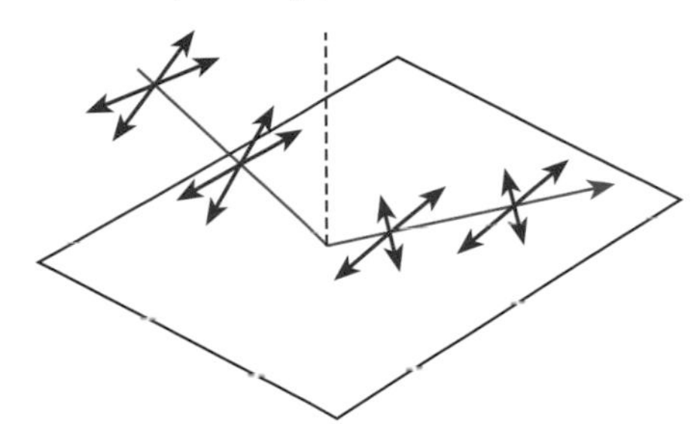

Brewster's angle

At one angle of reflection, called Brewster's angle, the light is plane polarised in a direction parallel to the reflecting surface.

The particular angle at which complete polarisation takes place depends on the refractive index, n, of the reflecting surface:

$$\tan \theta = n$$

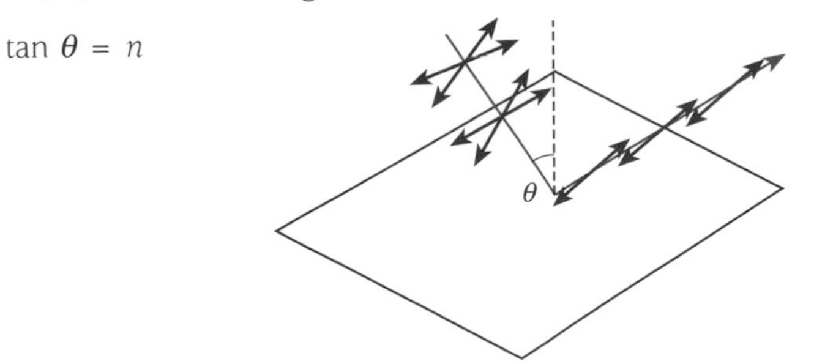

11.3 Polarisation of scattered light

When light is scattered by small particles, molecules, or small dust particles, the scattered light is plane polarised in a direction perpendicular to the plane of the scattered light and the direction of propagation.

The blue sky is scattered light that is plane polarised.

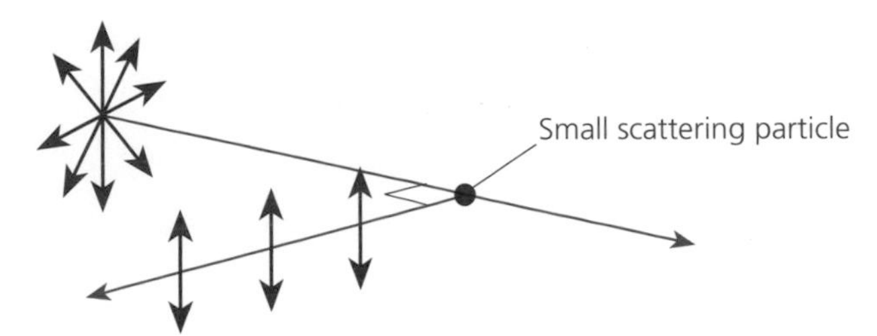

11.4 Detecting polarised light

The detection of polarised light requires the use of an analyser, which can be a Polaroid film or any other system that produces polarised light.

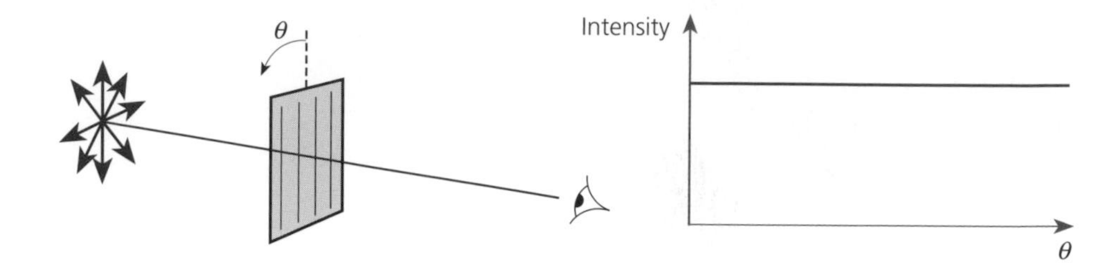

If the incident beam is unpolarised, as the analyser rotates, the intensity on the output side remains constant.

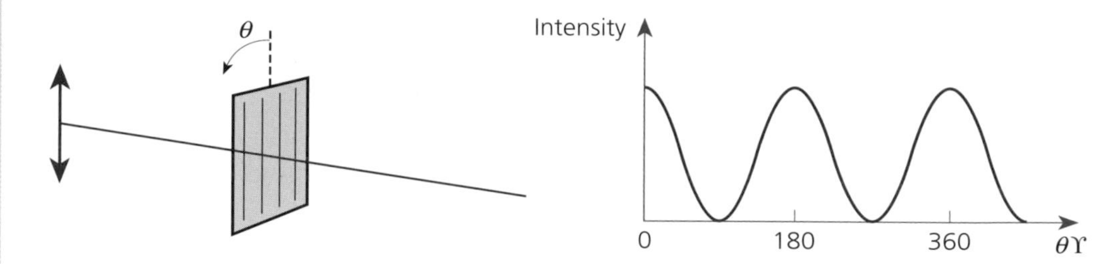

If the incident beam is polarised, as the analyser rotates, the intensity on the output side changes with θ. The intensity varies as $\cos^2 \theta$.

transmitted intensity = incident intensity × $\cos^2 \theta$

11.5 Circularly polarised light

This consists of two plane polarised beams at right angles, out of phase by $\frac{\pi}{2}$. When these two oscillations are added together, a polarised wave is generated which rotates with time.

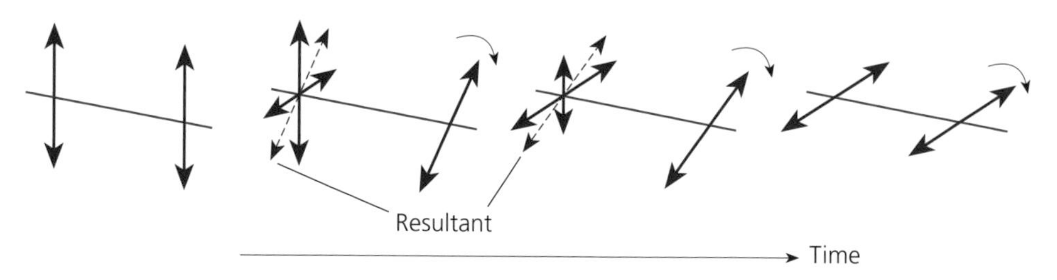

Worked examples

Q1 Calculate Brewster's angle for light reflected from glass of refractive index 1.5.

$\tan \theta = n$ $\tan \theta = 1.5$ $\theta = 56.3°$

Q2 Unpolarised light passes through two Polaroid films, which are at an angle of 30° to each other. What is the percentage reduction in intensity?

$I_{out} = I_{in} \cos^2 30$ $I_{out} = I_{in} \times 0.75$

There is a 25% reduction in intensity.

Q3 Incident light falls on a Polaroid sheet, which reduces the intensity by 15%. The partially polarised light then travels through a second sheet at the same angle. What is the percentage reduction in intensity?

Let the initial intensity = 1.

New intensity after first Polaroid = 0.85

After the second Polaroid = $0.85 \times 0.85 = 0.72$

Reduction in intensity = 28%

You should now know:

- **how to produce polarised light by transmission, reflection and scattering**
- **Brewster's angle**
- **how to use an analyser to detect polarised light**
- **how to describe circularly polarised light**

TOPIC 8 Fields

1 Fields of force

A field of force is said to exist in a region of space when a suitable object placed in the region experiences a force. We shall consider three fields of force:

1. **Gravitational field — in which the object is a mass that experiences the force.**
2. **Electric field — in which the object is a charge that experiences the force.**
3. **Magnetic field — in which the object is a moving charge or a magnet that experiences the force.**

Remember the experiment in which iron filings are sprinkled on paper above a bar magnet? The filings align themselves along the field lines.

Drawing fields of force

A field is a vector quantity and has both a magnitude and a direction. We represent fields by drawing lines of force:

- the density of the lines indicates the magnitude of the field
- the direction of the lines, indicated by arrows, shows the direction of the force on a suitably defined object in the region of space

Magnets with large field strengths exert large forces on each other.

1.1 Field strength

The magnitude of all fields, the field strength, is obtained by placing a suitable object in the field and measuring the force. The object must be small enough to have no effect on the field being measured.

Uniform fields

In uniform fields, the force is constant at all points and is shown by lines of force that are parallel.

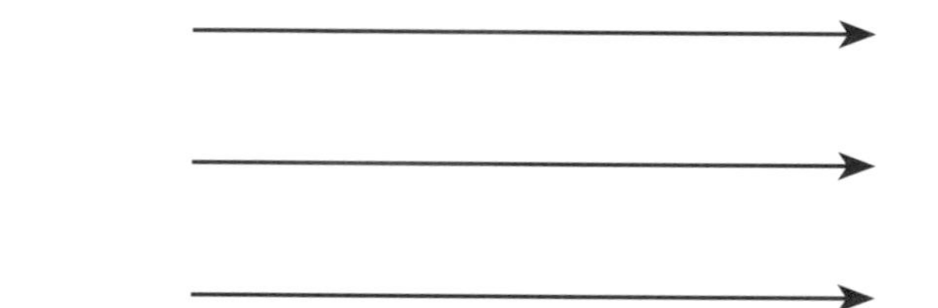

Non-uniform fields

In non-uniform fields, the lines of force change in density as the strength of the field changes.

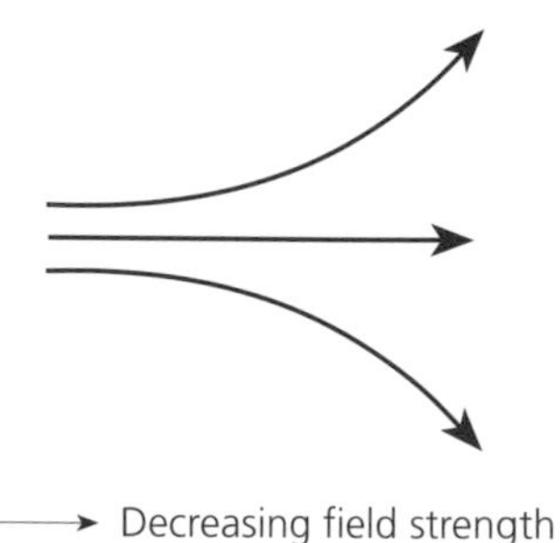

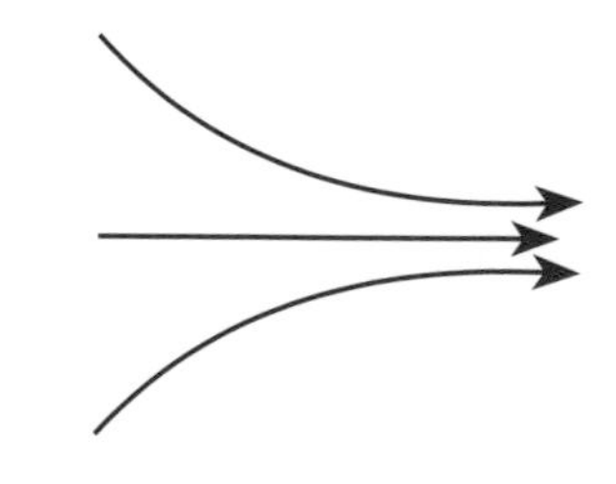

1.2 Change in potential energy in fields of force

When a suitable object is in a field, the object experiences a force. If the object is moved in the field, then energy is either released or supplied, depending on the direction of motion.

- **When the object is moved against the field, energy must be supplied because work is done on the object to move it against the field. There is an *increase* in the potential energy of the object.**
- **When the object is moved in the direction of the field, energy is released as the object does work. There is a *decrease* in the potential energy of the object.**

Change in potential when moving at an angle to a field

When an object moves at an angle to a field line, in order to calculate the potential we must calculate the force on the object along its direction of motion.

This is exactly the same concept as calculating the work done when moving a mass by a force at an angle to the direction of travel.

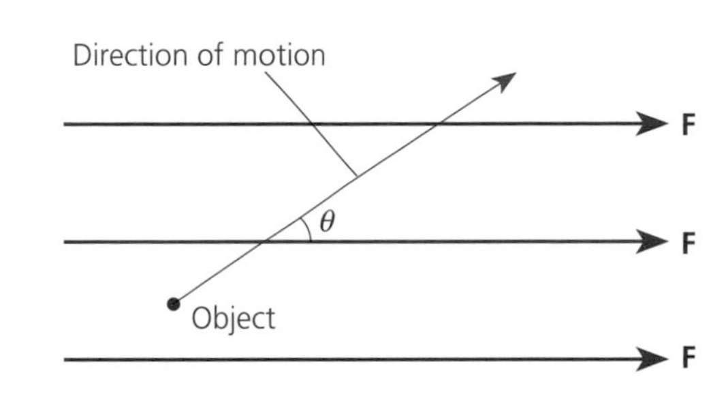

force in the direction of motion = **F** cos θ

1.3 Absolute potential at a point in a field

In the previous section, we indicated the change in potential as the object moves in the field. It is possible to assign an absolute value to the potential at a point in a field, provided a zero is chosen for the potential.

This is usually taken as being when the object is at infinity. The absolute potential at a point in the field is then the work done in moving the object from infinity to the point. Depending on the system being considered, the absolute potential may be positive or negative.

We have already stated that the potential energy of a mass *m* at a height *h* above the ground is *mgh*.

Other zeros of potential energy

In some calculations, it is convenient to choose an arbitrary zero for potential energy calculations, as in the case of motion under gravity on the Earth, where it is often convenient to choose the surface of the Earth as an arbitrary zero.

Equipotential lines

Since energy changes as objects move along field lines, lines of equal potential, known as equipotential lines, are always at right angles to the field lines.

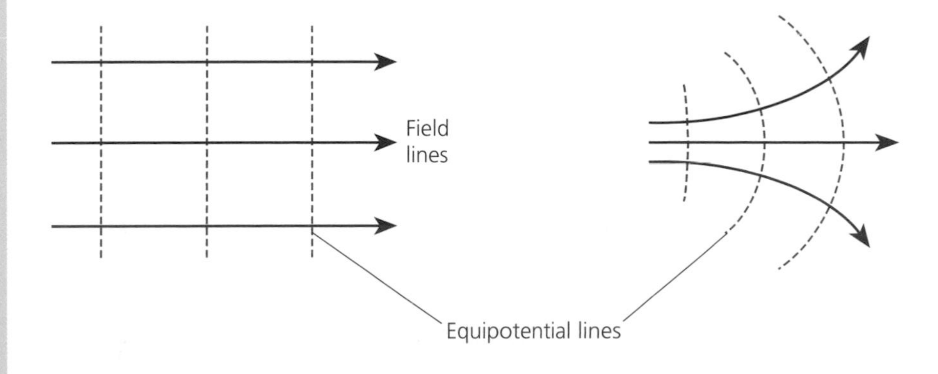

Worked examples

Q1 Calculate the change in potential when an object moves at right angles to a field line.

Since $\theta = 90°$, the force along the direction of travel = **F** cos θ = **F** cos 90 = 0

change in potential = 0

Q2 Sketch the equipotential lines above the surface of the Earth.

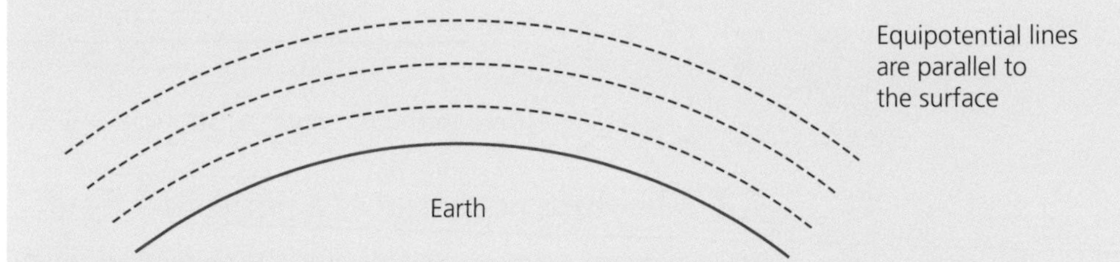

Q3 Sketch the equipotential lines created by the following field pattern.

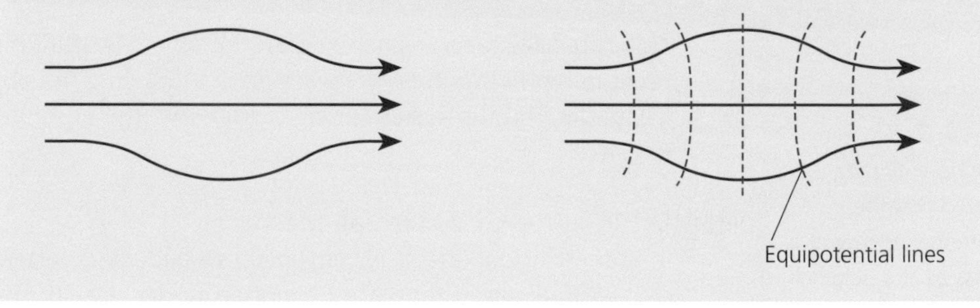

You should now know:

- some general ideas that apply to all fields, and the general definition of field strength
- the method used to draw field patterns
- about equipotential lines and their relationship to fields
- how to calculate the change in potential of an object moving in a field
- the definition of a zero of potential, and absolute potential

2 *Gravitational fields*

A gravitational field at a point is detected by observing the force on a mass placed at the point.

Gravitational fields have the following characteristics:

- they are generated by objects that have mass
- they are always attractive
- they are the weakest of the three fields we shall consider

Gravitational field strength g at a point is the force per unit mass placed at the point. It is measured in units N kg^{-1}.

In the ideal case, the mass should be as small as possible so as not to affect the field being measured.

$$g = \frac{\text{gravitational force}}{\text{mass}} = \frac{F}{m} \text{ N kg}^{-1}$$

Since the force is always attractive, the field lines around a spherical mass M are radial.

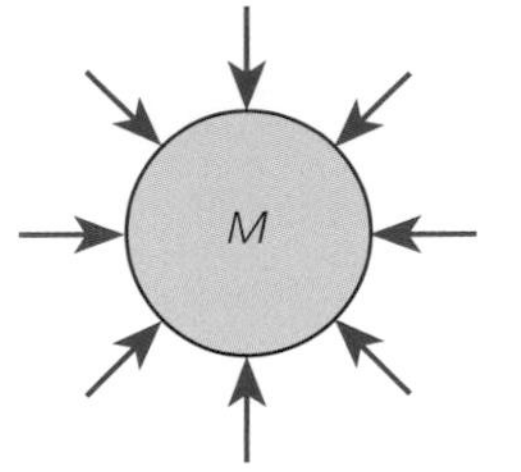

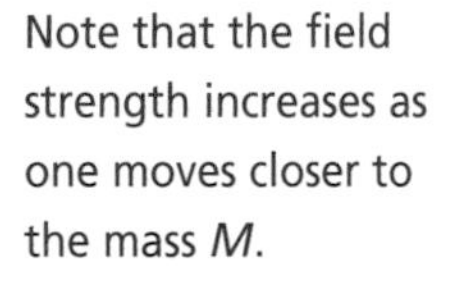

Note that the field strength increases as one moves closer to the mass M.

2.1 Newton's law of gravitation

If two masses m_1 and m_2 are separated by a distance d, the force of attraction of one mass on the other is:

$$F \propto \frac{m_1 m_2}{d^2} = \frac{G m_1 m_2}{d^2}$$ This is **Newton's law of gravitation.**

$$G = \text{gravitational constant} = 6.67 \times 10^{-11} \text{ N m}^2 \text{ kg}^{-2}$$

The law assumes that the masses concerned are point masses. For spheres of uniform density, outside the spheres the same formula can be applied and it can be assumed that the mass of the sphere is concentrated at the centre.

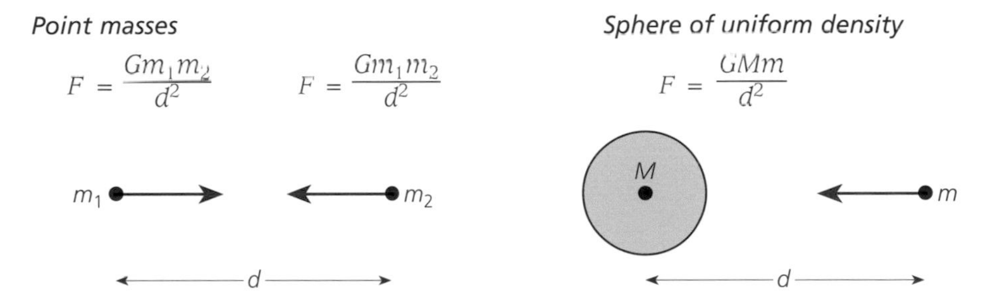

2.2 Gravitational field strength and Newton's law

When a mass m is placed in a gravitational field, $g = \frac{F}{m}$

If the mass m is in a gravitational field produced by a mass M:

$$\text{force} = \frac{GMm}{d^2}, \text{ hence } g = \frac{GMm}{md^2} = \frac{GM}{d^2}$$

A mass on the surface of the Earth

$$g = \frac{F}{m} = \frac{GmM_E}{mr_E^2} = \frac{GM_E}{r_E^2}$$

where M_E is the mass of the Earth and r_E is the radius of the Earth.

2.3 The acceleration due to gravity and *g* the gravitational field strength

Newton's second law of motion states that:

$$\text{force} = \text{mass} \times \text{acceleration}$$

A mass m on the surface of the Earth experiences a gravitational force, $\frac{GmM_E}{r_E^2}$ = $m \times$ acceleration due to gravity, which will cause the mass to accelerate towards the Earth.

Although these two are numerically the same, they are different quantities.

$$\textbf{acceleration due to gravity} = \frac{GM_E}{r_E^2} = \textbf{gravitational field strength} = g$$

The acceleration is independent of m, so all objects on the surface of the Earth accelerate at the same rate towards the surface.

acceleration due to gravity (9.81 m s^{-2}) = gravitational field strength (9.81 N kg^{-1})

2.4 Mass and weight

The mass of a body is a measure of its inertia, i.e. its reluctance to accelerate when a force is applied. The mass of a body is constant.

The weight of the same body is the force that gravity exerts on it. On the Earth, a mass of 1 kg has a weight = mass×acceleration = 1×9.81 = 9.81 N. The weight of an object varies depending on the magnitude of the gravitational field in which it is placed.

Mass and weight are often confused since, in order to compare masses, we often compare the forces that gravity exerts on them. Two objects of the same mass in the same gravitational field have the same weight.

Worked examples

Q1 The mass of the Earth is 5.98×10^{24} kg and it has a radius of 6.38×10^6 m. A small communication satellite of mass 20 kg is placed in orbit 4.23×10^7 m above the surface. Calculate the field strength at this height and the force on the satellite.

$$g = \frac{GM^E}{d^2} \qquad d = 6.38 \times 10^6 + 4.23 \times 10^7 = 4.87 \times 10^7 \text{m}$$

$$g = \frac{6.67 \times 10^{-11} \times 5.98 \times 10^{24}}{(4.87 \times 10^7)^2} = 1.69 \times 10^{-1}\,\text{N kg}^{-1}$$

force on the satellite = $gm = 1.69 \times 10^{-1} \times 20 = 3.36$ N

Q2 The Moon has a mass of 7.35×10^{22} kg and a radius of 1.74×10^6 m. What is the gravitational field strength on the surface of the Moon and its ratio with that on the Earth?

$$g = \frac{GM_M}{r_M^2} = \frac{6.67 \times 10^{-11} \times 7.35 \times 10^{22}}{(1.74 \times 10^6)^2} = 1.62\,\text{N kg}^{-1}$$

$$\frac{g_M}{g_E} = \frac{1.62}{9.81} = 0.17$$

Q3 What is the weight of a 1 kg mass on the Sun? The Sun has a mass of 1.99×10^{30} kg and a radius of 6.96×10^8 m.

$$F = \frac{Gm_1m_2}{d^2} = \frac{6.67 \times 10^{-11} \times 1 \times 1.99 \times 10^{30}}{(6.96 \times 10^8)^2} = 274\,\text{N}$$

You should now know:

- **the definition of gravitational field strength**
- **Newton's law of gravitation**
- **the relationship between field strength and Newton's law**
- **the relationship between field strength and acceleration due to gravity**
- **the difference between mass and weight**

3 Circular motion in the solar system — satellites

When an object moves in a circle, a force, called the centripetal force, acts on the object. Hence there is an acceleration towards the centre of the circle.

In the solar system, when the Earth moves around the Sun, the centripetal force is provided by the gravitational attraction between the Sun and the Earth. Using the ideas of circular motion, the orbital velocity of the Earth about the Sun can be calculated.

Note that the orbital velocity does not depend on the mass of the orbiting object.

$$F = \frac{GM_SM_E}{R^2} = \text{centripetal force} = \frac{M_Ev^2}{R} = M_ER\omega^2$$

$$\frac{GM_SM_E}{R^2} = \frac{M_Ev^2}{R}$$

$$v^2 = \frac{GM_S}{R}$$

$$\boldsymbol{v = \sqrt{\frac{GM_S}{R}}}$$

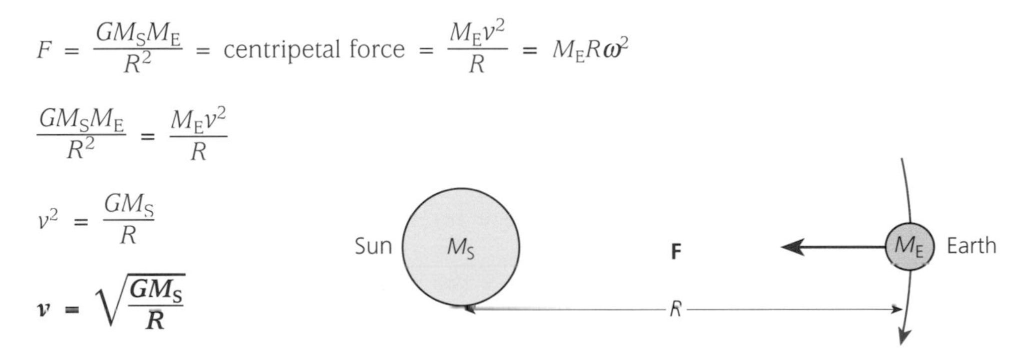

3.1 Kepler's law and the period of rotation of planets

Using the above:

$$\frac{GM_SM_E}{R^2} = M_ER\omega^2 \qquad \text{but } T = \frac{2\pi}{\omega}$$

$$\frac{GM_SM_E}{R^2} = \frac{M_ER4\pi^2}{T^2}$$

$$GM_ST^2 = 4\pi^2R^3$$

$$\boldsymbol{T^2 \propto R^3}$$

The (period of rotation of the planet)^2 is proportional to the (mean radius of the orbit)^3. This is Kepler's law.

3.2 Satellites

These same equations can be applied to satellites orbiting the Earth, except that the mass at the centre is now the Earth, not the Sun.

Geostationary orbits

It is possible to place satellites a suitable distance above the Earth, orbiting in the plane of the equator, such that the period of rotation is the same as that of the Earth. These satellites will always stay above the same place on the surface of the Earth and are said to be in geostationary orbits.

Communication satellites are almost always placed in geostationary orbits.

3.3 Gravitational potential

A mass m in a gravitational field has potential energy, since work must be done on the mass to move it against the field. The potential energy increases when it moves against the field, and decreases when it moves with the field.

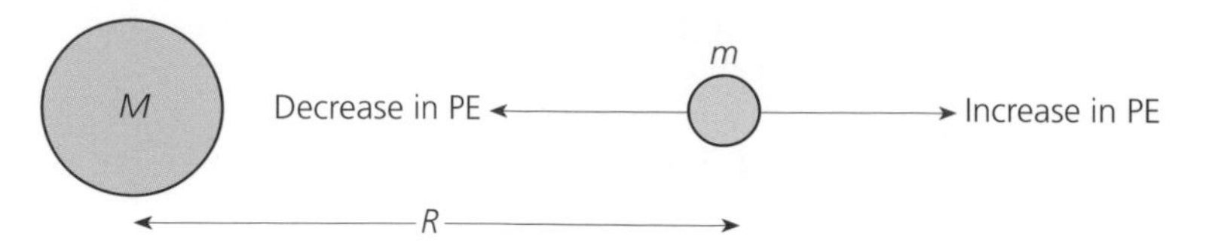

The zero of potential is chosen to be at $R = \infty$, and so the potential is negative as we move closer to the mass M. You will not need to prove this; just know the form of the equation.

The gravitational potential at a point in a gravitational field is defined as the work done in bringing a mass of 1 kg from infinity to the point. Using Newton's law of gravitation, it is possible to show that the gravitational potential V at a distance R away from an object of mass M is:

$$V = -\frac{GM}{R}$$

3.4 Gravitational field strength and potential gradient

The minus sign is inserted because the gravitational field acts in the opposite direction to the distance moved.

If a 1 kg mass on the surface of the Earth is moved a distance 1 m away from the Earth, work is done and the gravitational potential increases.

$$\text{work done} = \text{increase in potential} = \Delta V = \text{force} \times \text{distance} = -g \times \Delta x$$

$$g = -\frac{\Delta V}{\Delta x} = \text{the potential gradient}$$

Gravitational field strength is numerically equal to the **gravitational potential gradient**.

3.5 PE and KE of satellites

Note that as R gets smaller, the kinetic energy increases and the orbital velocity increases.

A satellite mass m in an orbit at a radius R has a potential energy $E_p = -\frac{GM_Em}{R}$

The kinetic energy of the satellite $= \frac{1}{2}mv^2$, but $v^2 = \frac{GM_E}{R}$, so $E_k = \frac{1}{2}\frac{GM_Em}{R}$

Total energy of the satellite

As the satellite moves closer to the Earth, it loses potential energy. Half goes to increase the kinetic energy, the other half is lost in friction.

$$\text{total energy} = \text{PE} + \text{KE} = -\frac{GM_Em}{R} + \frac{1}{2}\frac{GM_Em}{R} = -\frac{1}{2}\frac{GM_Em}{R}$$

3.6 Escape velocity

To escape from the Earth, the initial kinetic energy of the satellite must be greater than the gain in potential energy as the satellite moves from the surface to an infinite distance away.

$$\text{gain in potential energy} = 0 - \left(-\frac{GM_Em}{R}\right) = \frac{GM_Em}{R}$$

PE at ∞∞∞ PE on the surface

The orbital velocity of a satellite is: $v = \sqrt{\frac{GM_E}{R}}$ but $g = \frac{GM_E}{R^2}$ so $v = \sqrt{gR}$ so the escape velocity is $\sqrt{2}$ times the orbital velocity.

$$\text{initial kinetic energy} = \frac{1}{2}mv^2 = \text{gain in PE} = \frac{GM_Em}{R} \qquad v^2 = \frac{2GM_E}{R}$$

$$\textbf{escape velocity } v = \sqrt{\frac{2GM_E}{R}} = \sqrt{2gR}$$

Worked examples

Q1 Calculate the period of rotation of the Moon about the Earth in days. The Moon orbits the Earth at a radius of 3.85×10^8 m, and the Earth has a mass of 5.98×10^{24} kg.

$$GM_ET^2 = 4\pi^2R^3$$

$$T^2 = \frac{4 \times \pi^2 \times (3.85 \times 10^8)^3}{6.67 \times 10^{-11} \times 5.98 \times 10^{24}} = 5.60 \times 10^{12}$$

$$T = 2.38 \times 10^6 \text{ s} = 27.5 \text{ days.}$$

Q2 Calculate the change in gravitational potential when a satellite of mass 20 kg is placed in an orbit 200 km above the Earth. The radius of the Earth is 6.38×10^6 m.

$$\text{change in potential} = -\frac{GMm}{R_1} - \left(-\frac{GMm}{R_2}\right)$$

$$= 6.67 \times 10^{-11} \times 5.98 \times 10^{24} \times 20\left(-\frac{1}{6.58 \times 10^6} + \frac{1}{6.38 \times 10^6}\right)$$

$$= 3.8 \times 10^7 \text{ J}$$

Q3 Calculate the escape velocity of the lunar space capsule from the surface of the Moon. The mass of the Moon is 7.35×10^{22} kg and its radius is 1.74×10^6 m.

$$v^2 = \frac{2GM}{R} = \frac{2 \times 6.67 \times 10^{-11} \times 7.35 \times 10^{22}}{1.74 \times 10^6}$$

$$v^2 = 5.64 \times 10^6 \qquad v = 2370 \text{ m s}^{-1}$$

Remember, g on the Moon is 1.62 N kg^{-1}.

$$\text{or } v = \sqrt{2Rg} = \sqrt{2 \times 1.74 \times 10^6 \times 1.62} = 2370 \text{ m s}^{-1}$$

You should now know:

- **Kepler's law and its derivation**
- **the definition of geostationary satellite orbits**
- **gravitational potential and its relationship to the gravitational field strength**
- **the potential and kinetic energy of satellites and how to calculate the escape velocity**

4 Electric fields

4.1 Force between electrical charges

There are two types of electrical charges: positive and negative. When separated by a distance, charges experience forces: like charges repel, unlike charges attract.

F F

An electric field is detected by observing the force on a charge placed at a point.

Remember, gravitational fields are only attractive.

Electric fields have the following characteristics:

- they are generated by stationary electric charges
- they can be attractive or repulsive
- they are very strong fields

The **electric field strength *E*** at a point is the force per unit positive charge placed at the point. The direction of the field is the direction of the force on the positive charge.

$$E = \frac{F}{Q}$$ Units, newtons coulomb^{-1} (N C^{-1})

In the ideal case, the charge should be as small as possible so as not to affect the field it is measuring.

4.2 Uniform electric fields

A charge Q coulombs in an electric field experiences a force $Q \times E$ newtons.

Positive charge

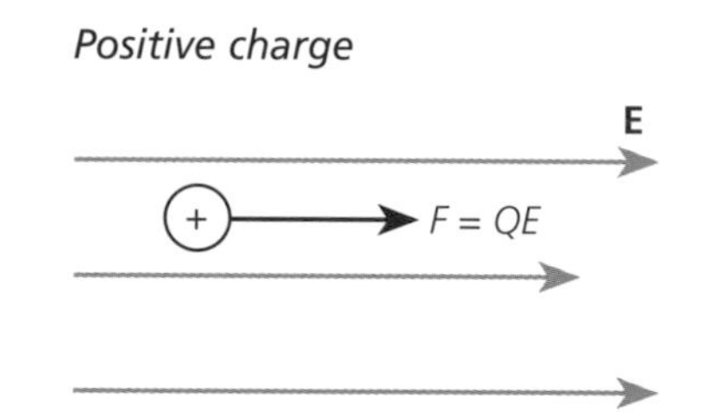

The force is in the same direction as E and has magnitude $Q \times E$.

Negative charge

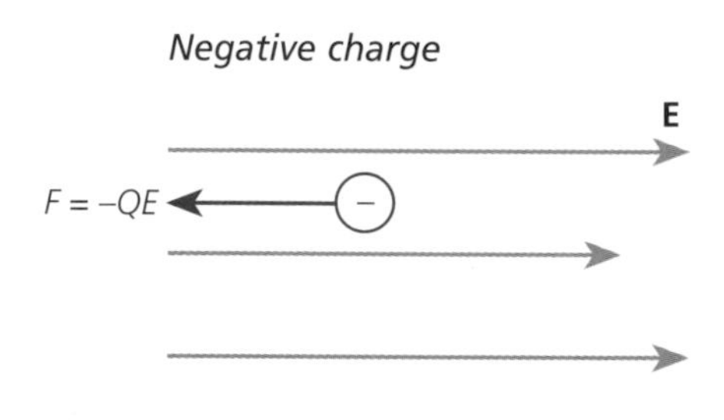

The force is in the opposite direction to E and has magnitude $Q \times E$.

Uniform electric fields are used to deflect charges, as in the oscilloscope. The uniform field is produced by placing opposite charges on the two plates.

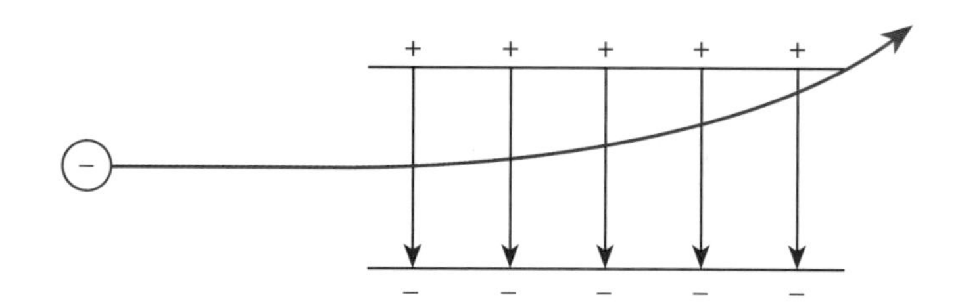

4.3 Coulomb's law

When two point charges Q_1 and Q_2 are separated by a distance d, the force on one charge due to the other is:

$$F = \frac{1}{4\pi\varepsilon_0}\frac{Q_1 Q_2}{d^2}$$ This is **Coulomb's law.**

ε_0 is a constant and represents the permittivity of free space $= 8.85 \times 10^{-12}$ C^2 N^{-1} m^{-2}.

4.4 Electric field due to a point charge

To obtain the magnitude of the electric field at a point d away from a point charge $+Q$, we place a small charge $+q$ at the point and measure the force.

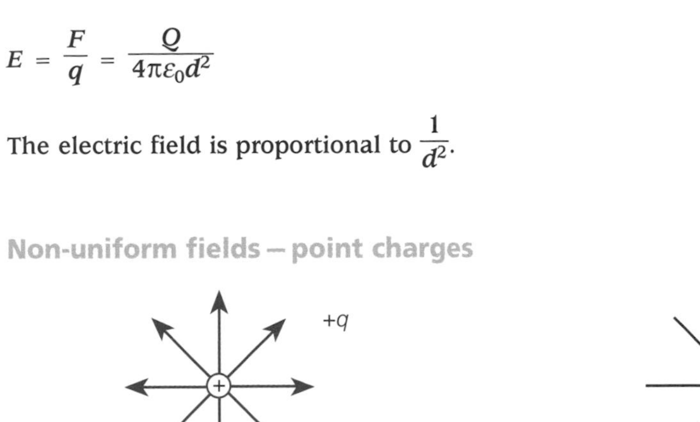

$$F = \frac{qQ}{4\pi\varepsilon_0 d^2}$$

$$E = \frac{F}{q} = \frac{Q}{4\pi\varepsilon_0 d^2}$$

The electric field is proportional to $\frac{1}{d^2}$.

Non-uniform fields – point charges

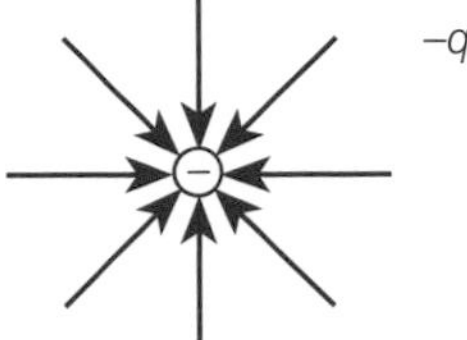

Other non-uniform fields

Electric dipole

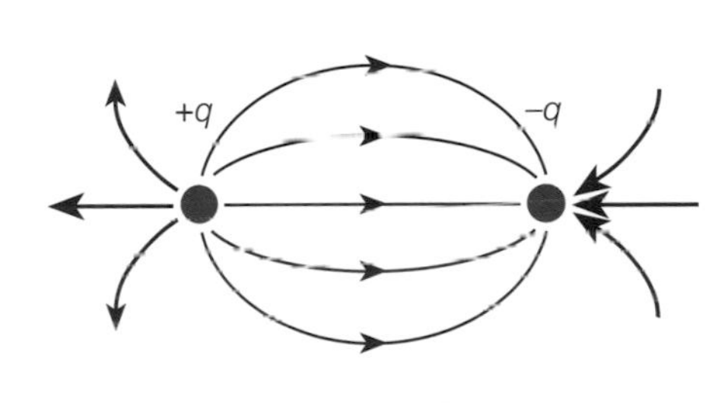

Charged hollow sphere

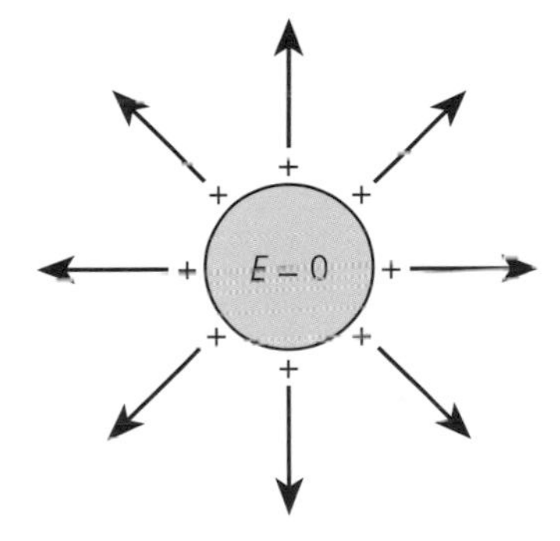

This is used in electrical circuits to isolate components from outside electrical interference.

For the charged hollow sphere, outside the sphere the field is the same as if all the charge were at the centre of the sphere. There is no electric field inside.

Electric field due to a flat conducting sheet

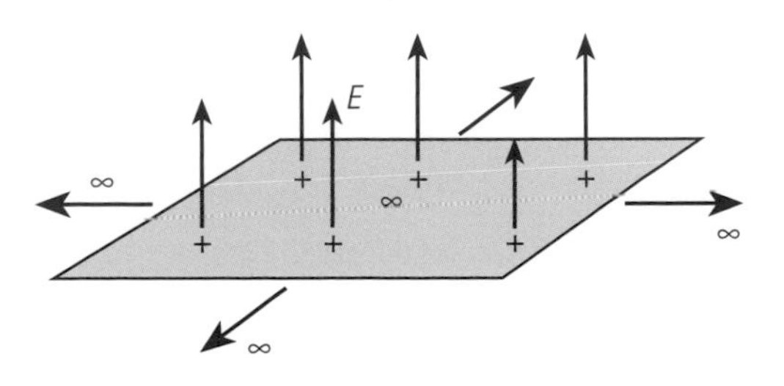

With an infinite sheet of charge, the electric field is perpendicular to the surface and has a magnitude:

$$E = \frac{\sigma}{\varepsilon} \qquad \sigma = \text{the charge per unit area of the sheet}$$

Worked examples

Q1 Calculate the electric field 20 cm away from a point charge of 1.6×10^{-8} C.

$$E = \frac{Q}{4\pi\varepsilon_0 d^2} = \frac{1.6 \times 10^{-8}}{4 \times \pi \times 8.85 \times 10^{-12} \times (0.2)^2} = 3.6 \times 10^3 \text{ N C}^{-1}$$

Q2 Three charges lie in line as shown. What is the force on the middle charge?

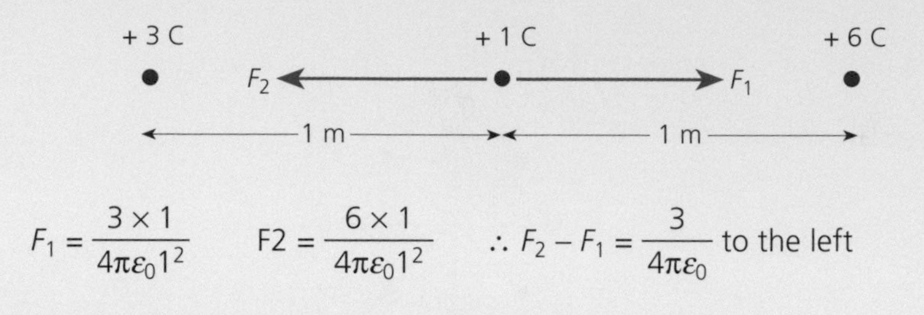

$$F_1 = \frac{3 \times 1}{4\pi\varepsilon_0 1^2} \qquad F2 = \frac{6 \times 1}{4\pi\varepsilon_0 1^2} \qquad \therefore F_2 - F_1 = \frac{3}{4\pi\varepsilon_0} \text{ to the left}$$

Q3 Two metal plates 20 cm² in area carry a charge of 1.6×10^{-12} C. Calculate the electric field between the plates.

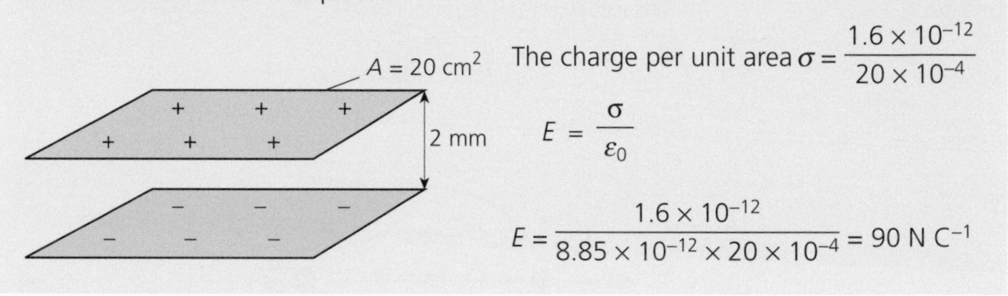

The charge per unit area $\sigma = \frac{1.6 \times 10^{-12}}{20 \times 10^{-4}}$

$$E = \frac{\sigma}{\varepsilon_0}$$

$$E = \frac{1.6 \times 10^{-12}}{8.85 \times 10^{-12} \times 20 \times 10^{-4}} = 90 \text{ N C}^{-1}$$

Do not fall into the trap of thinking that because there are two sheets of charge the field is double.

You should now know:

- the shape of electric fields around charges
- the definition of electric field strength
- Coulomb's law and its application
- the shape and magnitude of an electric field due to a sheet of charge

5 *Electric potential*

The electric potential at a point in an electric field is the work done per unit charge in bringing the charge from infinity to the point.

$$V = \frac{\text{work done}}{\text{charge}} = \frac{W}{Q}$$

Units, joules coulomb^{-1} (J C^{-1}) or volts

1 J C^{-1} = 1 V

In gravitational fields, the potential at a point is always negative, since gravitational forces are always attractive and energy is given out when moving from infinity to the point.

In the case of electrical fields, the potential at a point can be negative or positive.

(+) A ∞ (+) F

The potential here will be positive, as work must done on the charge (energy supplied) to move the charge from infinity to the point A.

(−) A ∞ (+) F

The potential here will be negative, as the charge does work (energy is released) when it is moved from infinity to the point A.

5.1 Potential due to a point charge

Lines of equal potential form concentric spheres around the charge.

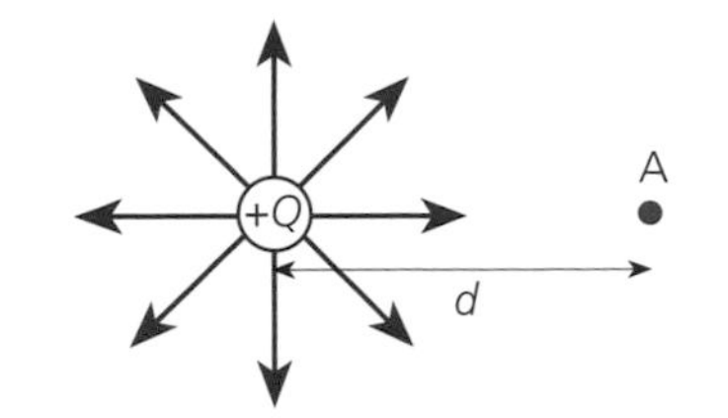

Potential V at a point A is given by:

$$V = \frac{Q}{4\pi\varepsilon_0 d}$$

5.2 Potential difference

This definition is equivalent to the one for potential difference in electrical circuits.

In many applications of electric fields, we are only interested in the potential difference when a charge moves from one position to another, i.e. the work done in moving between the two points.

The electric field and potential gradient – uniform fields

Consider a charge $+Q$ moving against a field E through a distance x.

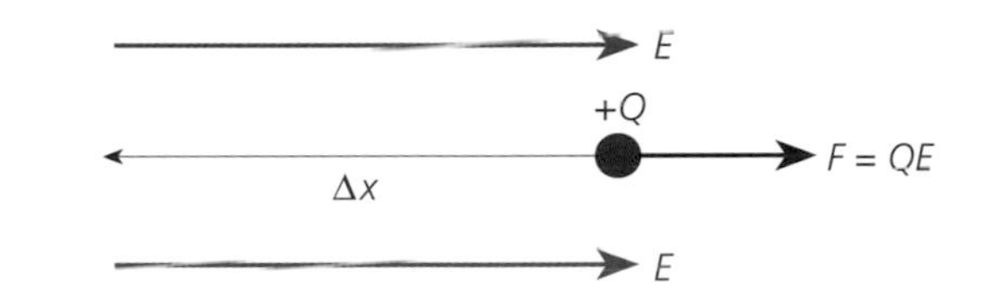

Minus sign because we are moving a distance x against the force.

Using this method, the units for electric field can also be expressed as V m^{-1} as well as N C^{-1}.

The work done in moving the charge Q a distance x against the field $= -Fx = -QE\Delta x$.

$$\text{work done per unit charge} = \text{change in potential } \Delta V = -\frac{QE\Delta x}{x} = -E\Delta x$$

$$\boldsymbol{E} = -\frac{\Delta \boldsymbol{V}}{\Delta \boldsymbol{x}}$$

The magnitude of the electric field equals the potential gradient.
The electric field is in the direction of decreasing electric potential.

Equipotential lines

As with equipotential lines in gravity, equipotential lines in electric fields are perpendicular to the direction of the field line.

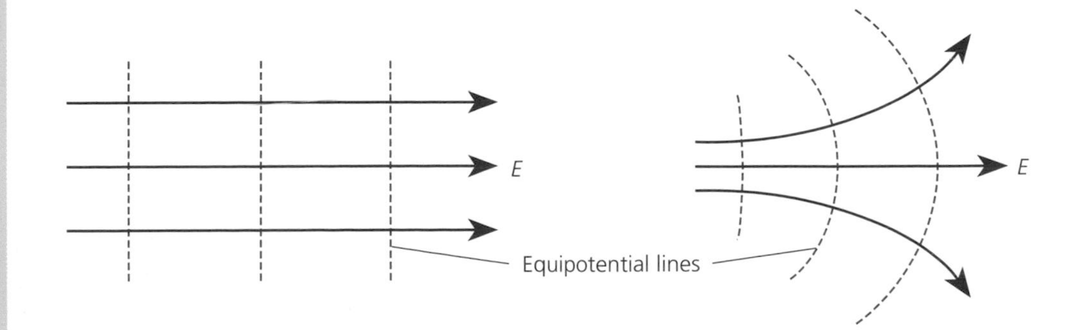

Remember to work in metres.

Worked examples

Q1 Two large metal plates separated by a distance of 10 cm carry opposite charges. The potential difference across the plates is 10 V. Calculate the electric field between the plates and the charge per unit area on the plates.

$$E = \frac{\Delta V}{\Delta x} = \frac{10}{0.1} = 100\,\text{V m}^{-1}$$

But $E = \dfrac{\sigma}{\varepsilon_0}$ $\quad \therefore \sigma = 100 \times 8.85 \times 10^{-12}$

$$\sigma = 8.85 \times 10^{-10}\,\text{C}$$

Q2 A positive charge of 1×10^{-6} C is moved in the field of a second positive charge of 5×10^{-4} C. The charge moves outwards from 1 m to 7 m. Calculate the change in electrical potential.

$$V = \frac{Q}{4\pi\varepsilon_0 d} = \frac{5 \times 10^{-4} \times 1 \times 10^{-6}}{4\pi\varepsilon_0 \times 7} - \frac{5 \times 10^{-4} \times 1 \times 10^{-6}}{4\pi\varepsilon_0 \times 1}$$

$$= \frac{5 \times 10^{-4} \times 1 \times 10^{-6}}{4 \times \pi \times 8.85 \times 10^{-12}}\left(\frac{1}{7} - 1\right)$$

$$= -3.9\,\text{J C}^{-1}$$

Q3 An electron enters the parallel plate system illustrated below with a velocity of 1 × 107 m s^{-1}. What is its vertical acceleration between the plates and the vertical velocity on leaving the plates?

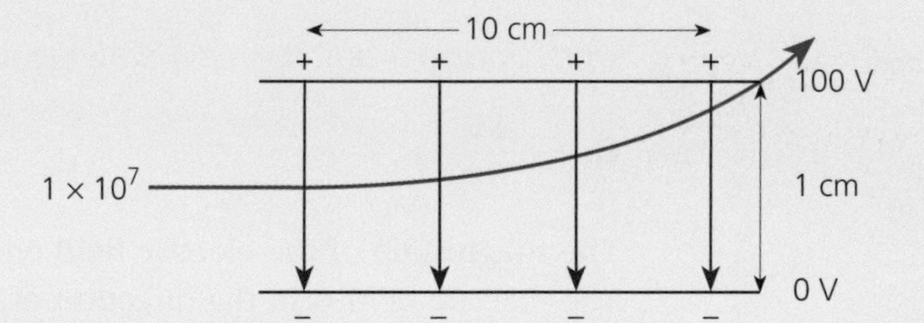

$$E \text{ between the plate} = \frac{\Delta V}{\Delta x} = \frac{100}{0.01} = 1 \times 10^{4}\,\text{V m}^{-1}$$

$$\text{time between the plate} = \frac{0.1}{1 \times 10^{7}} = 1 \times 10^{-8}\,\text{s}$$

vertical acceleration $F = qE = ma \quad 1.6 \times 10^{-19} \times 1 \times 10^{4} = 9.11 \times 10^{-31} \times a$

$$a = 1.76 \times 10^{15}\,\text{m s}^{-2}$$

Using $v = u + at$ in the vertical direction:

$v = 0 + 1.76 \times 10^{15} \times 1 \times 10^{-8}$

$v = 1.8 \times 10^{7}\,\text{m s}^{-1}$

You should now know:

- **the definition of electrical potential and potential difference**
- **the equation for the potential due to a point charge**
- **the relationship between potential gradient and electric field strength**
- **how to draw equipotential lines**

6 Magnetic fields

Magnetic fields are produced by moving charges or by currents in wires. In a simple bar magnet, there do not appear to be any currents, but the magnetic field is generated by electrons orbiting atoms that make up the structure of the magnet.

The direction of the magnetic field is defined in terms of the simple bar magnet, with field lines going from the north-seeking pole to the south-seeking pole.

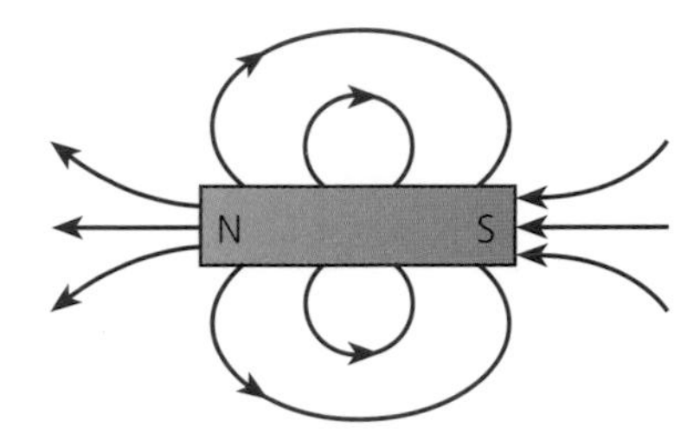

Note that single magnetic poles do not exist, so we cannot use similar definitions to those used for gravitational and electric field strengths.

6.1 Magnetic field strength

When a single wire, which carries a current I, is placed in a magnetic field, it experiences a force, and this is used to define the magnetic field strength.

The magnetic field strength B or the magnetic flux density is the force acting per unit current per unit length of wire placed at right angles to the field.

$B = \frac{F}{IL}$ **Units, tesla (T)**

The direction of the field is at right angles to the current and the force.

The directions of B, I and the force are given by the **left-hand motor rule**. Remember motors travel on the left in the UK.

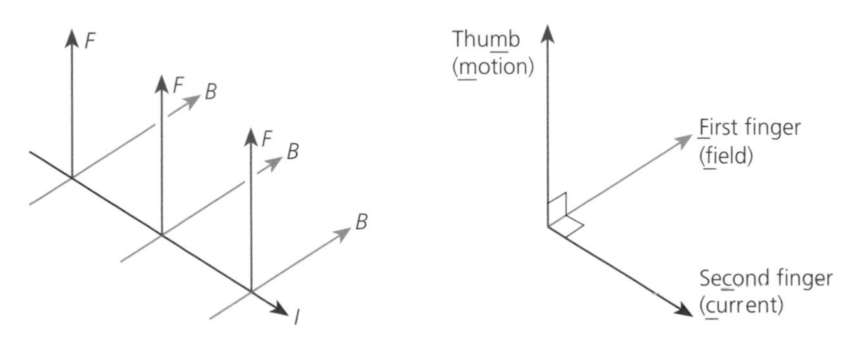

In the above, I is the conventional current, the direction of positive charge flow.

Another unit for B comes from

$B = \frac{\text{magnetic flux}}{\text{area}}$

measured in units weber per square metre (Wb m^{-2}). B is often called magnetic flux density.

Magnetic flux

When magnetic field lines pass through an area, we define a new quantity called the **magnetic flux**. Magnetic flux is of importance when we consider electromagnetic induction.

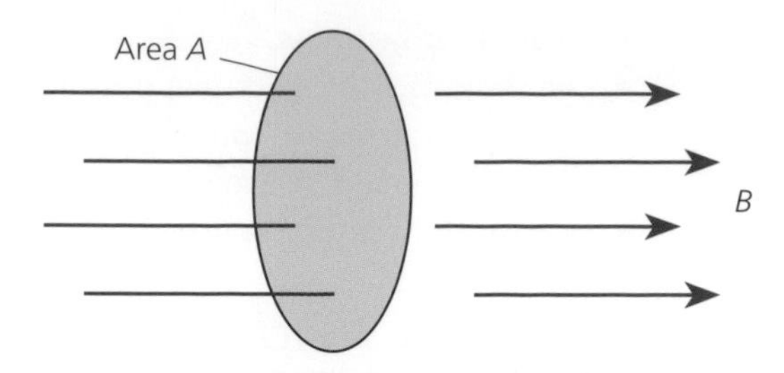

magnetic flux $= B \times A$ Units, weber (wb)

Magnetic field patterns

At the centre of a long solenoid, the magnetic field is uniform.

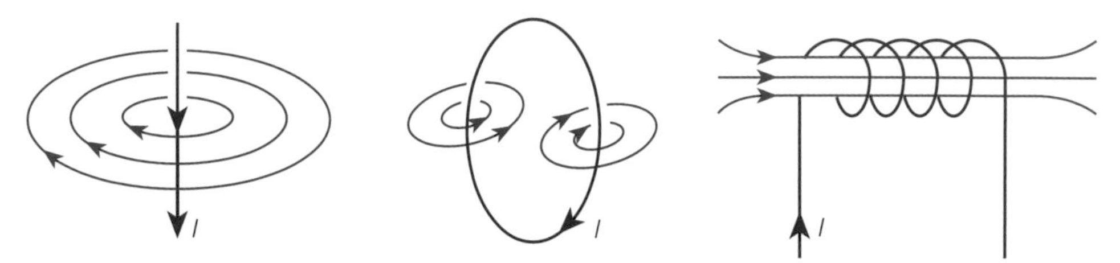

6.2 Force produced by a current I in a magnetic field

Using the definition of magnetic field strength, we can calculate the force generated by a straight wire of length L in a magnetic field.

When $\theta = 0$ the force is zero.

B and I perpendicular

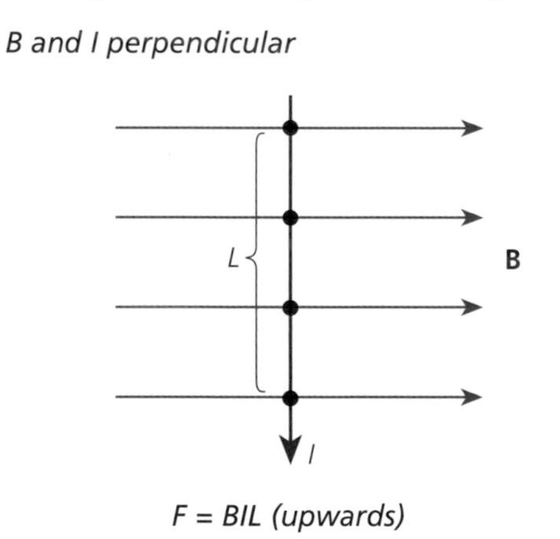

F = BIL (upwards)

B at an angle θ *to I*

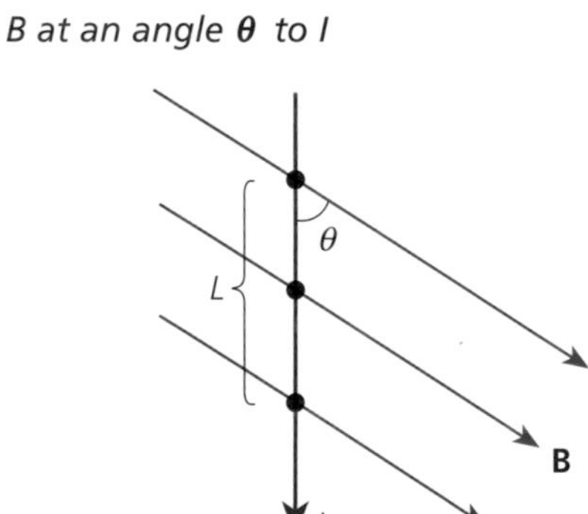

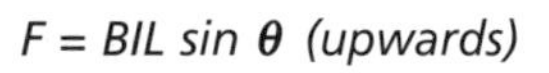

F = BIL sin θ *(upwards)*

Forces on coils

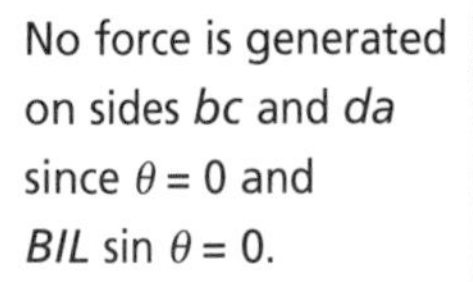

No force is generated on sides *bc* and *da* since $\theta = 0$ and $BIL \sin \theta = 0$.

The force F generated on each side $= BIL$.

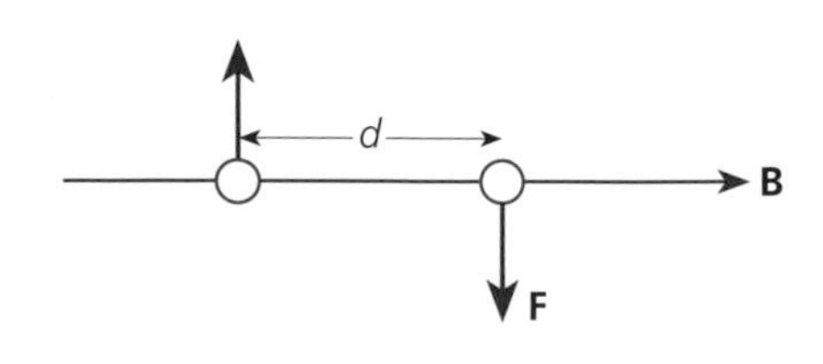

The torque generated is $F \times d$, which rotates the coil and is the basis of a d.c. motor.

6.3 Forces on moving charges

Current is rate of flow of charge, and the force on a wire in a magnetic field is generated by the moving charges.

I = rate of flow of charge = $\frac{q}{t}$

If the charge has a velocity v, in a time t it will travel a distance $L = vt$.

magnitude of the force = $BIL = B\frac{q}{t}vt = Bqv$

The direction of the force is given by the left-hand motor rule, remembering that the direction of the velocity is the same as the direction of the current, provided we are considering positive charges.

As the force is at right angles to v, the charges move in a circle. The force provided by the magnetic field is the centripetal force.

With negative charges such as electrons, the force is in the reverse direction.

The sign convention for showing the direction of fields perpendicular to the page is
⊗ B into page
⊙ B out of page.

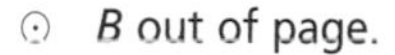

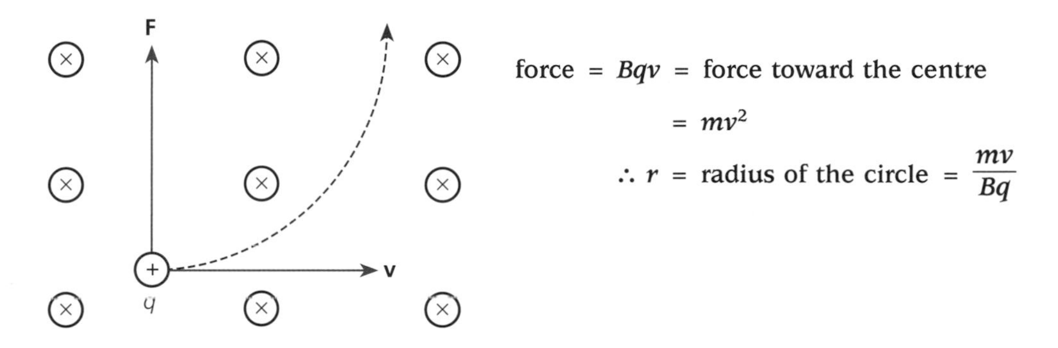

force = Bqv = force toward the centre

$= mv^2$

$\therefore r$ = radius of the circle = $\frac{mv}{Bq}$

Worked examples

Q1 A straight wire carries a current of 13 A. Calculate the force per unit length of the wire when it is placed perpendicular to a magnetic field of 0.2 T.

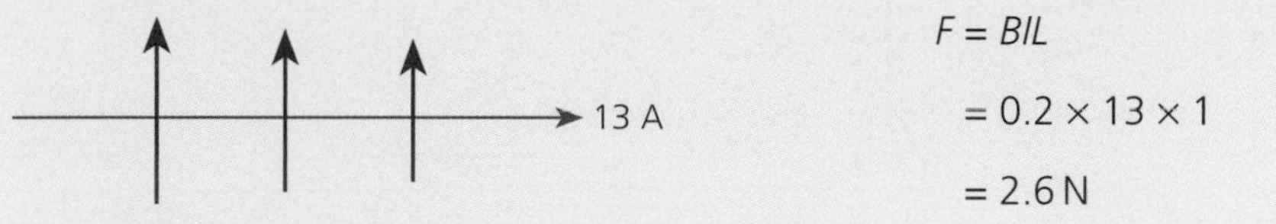

$F = BIL$

$= 0.2 \times 13 \times 1$

$= 2.6\,\text{N}$

Q2 The coil of a d.c. electric motor has dimensions 1 cm × 0.15 cm and 45 turns of wire and is in a magnetic field of 0.3 T. Calculate the maximum torque on the coil and the torque when it has rotated through 30°, when it carries a current of 20 mA.

current down one wire = 20 mA

force = $BIL = 0.3 \times 0.9 \times 0.01 = 2.7 \times 10^{-3}$ N

current down 45 wires = $45 \times 20 \times 10^{-3} = 0.9$ A

maximum torque = $F \times d = 2.7 \times 10^{-3} \times 0.0015 = 4.1 \times 10^{-6}$ N m

After rotation:

F is still 2.7×10^{-3} $\quad d' = d \cos 30$

torque = $F \times d' = 2.7 \times 10^{-3} \times 0.0015 \times \cos 30$
$= 3.6 \times 10^{-6}$ N m

Q3 An electron enters a magnetic field of 0.4 T with a velocity of 4.2 × 105 m s^{-1}. Calculate the force on the electron and the radius of the curved path it takes.

$$F = Bqv = 0.4 \times 1.6 \times 10^{-19} \times 4.2 \times 10^5 = 2.7 \times 10^{-14}\ \text{N}$$

$$r = \frac{mv}{Bq} = \frac{9.11 \times 10^{-31} \times 4.2 \times 10^5}{0.4 \times 1.6 \times 10^{-19}} = 6.0 \times 10^{-6}\ \text{m}$$

You should now know:

- **the definitions of magnetic field strength, magnetic flux density and magnetic flux**
- **the shape of the magnetic field patterns around a straight wire, coil and solenoid**
- **how to calculate the force on a current or a moving charge in a magnetic field**
- **how to calculate the radius of the circle around which a moving charge travels**

7 *Applications of electric and magnetic fields*

There are many examples of applications of electric and magnetic fields and their effects on moving charges.

7.1 Accelerating a charged particle in an electric field

In many systems (oscilloscope, mass spectrometer), charged particles need to acquire velocity, and the most common method is to use an electric field.

The acceleration of positive charges *The acceleration of negative charges*

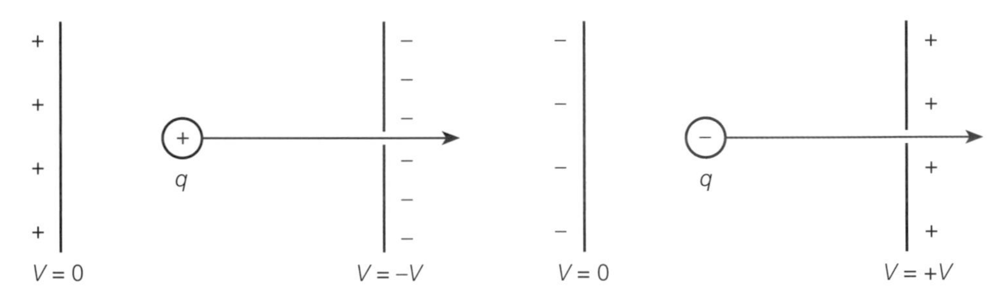

A uniform electric field E between the plates is produced by a potential difference V across the plates.

$$\text{force on the charge } q = qE = q\frac{V}{d} = ma$$

$$\text{acceleration } a = \frac{qV}{md}$$

$$\text{final velocity } v^2 = u^2 + 2as = 2 \times \frac{qV}{md} \times d = \frac{2qV}{m}$$

The final KE depends only on the charge and the voltage between the plates.

$$\textbf{gain in kinetic energy} = \tfrac{1}{2}mv^2 = \frac{1}{2}m\,\frac{2qV}{m} = qV$$

The electronvolt

We shall consider the energy gained in many physical situations by charges that are an integer multiple of the charge on an electron of 1.6×10^{-19} C. The energies gained are small, and it is convenient to introduce a new unit, the **electronvolt** (eV).

The electronvolt is a very small unit of energy. In nuclear physics calculations, we often use the unit MeV.
1 MeV = 1×10^6 eV

The electronvolt is the energy gained by an electron as it is accelerated through a potential difference of 1 volt.

$$1 \text{ eV} = 1.6 \times 10^{-19} \times 1 = 1.6 \times 10^{-19} \text{ J}$$

So, to convert from energy in electronvolts to energy in joules, multiply by 1.6×10^{-19}.

7.2 The motion of charges in combined electric and magnetic fields

If a magnetic and an electric field are placed at right angles to each other and in the correct direction, it is possible to generate forces on moving charges that cancel.

Note that in the case of negative charges, the electric and magnetic forces are reversed.

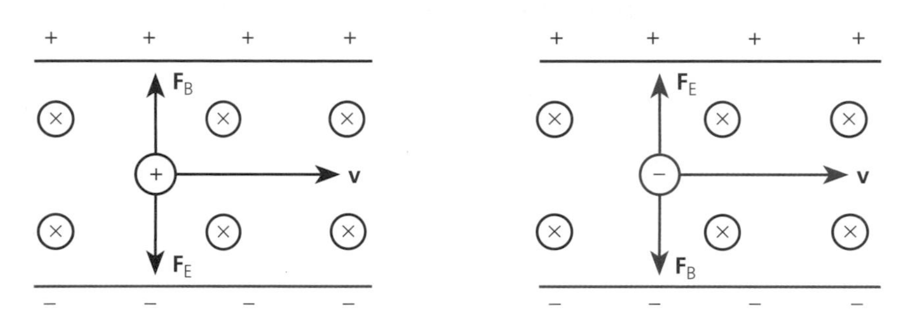

In both cases, the forces cancel when the charged particle has a certain velocity.

Note that this does not depend on the mass of the particle.

$$F_B = F_E$$

$$Bqv = Eq$$

when the velocity $v = \frac{E}{B}$

If particles enter the crossed field with a range of velocities, only those with a velocity $= \frac{E}{B}$ pass through undeflected. This is called a **velocity selector**.

The directions would be reversed for negative charges.

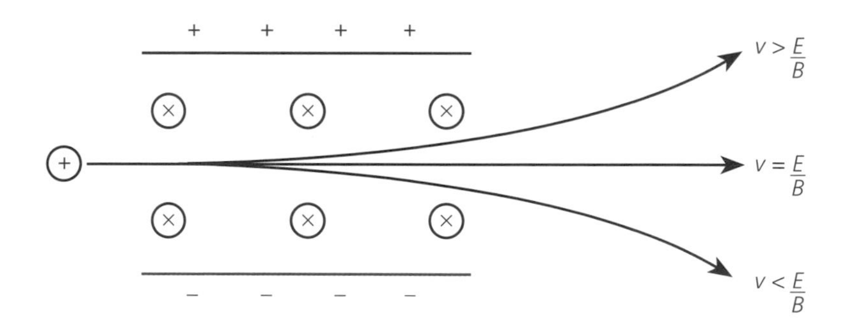

7.3 Applications

The mass spectrometer

Magnetic fields can be used in conjunction with velocity selectors to measure the masses of different atoms in a substance. In its simplest form, high-speed ions — atoms that have lost one electron — pass into a velocity selector. Leaving the selector, all the ions have the same velocity. They then pass into a magnetic field and travel circular paths with different diameters.

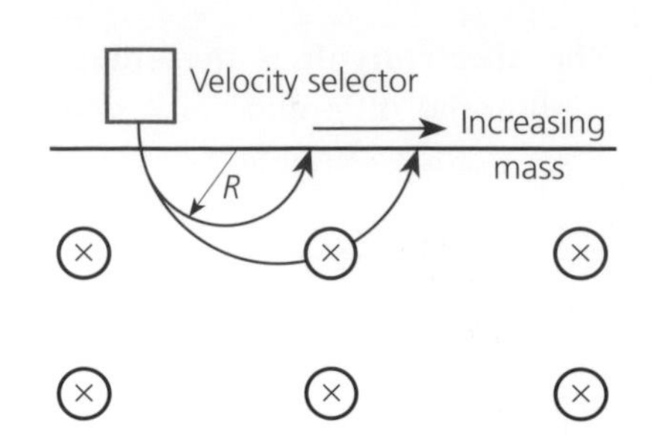

In the magnetic field:

$$Bqv = \frac{mv^2}{R}$$

$$m = \frac{BqR}{v}$$

$$m \propto R$$

The radius of the path is proportional to the mass of the ions, provided they enter the field with the same velocity.

The cyclotron

This is used to accelerate charged particles making circular orbits in a magnetic field.

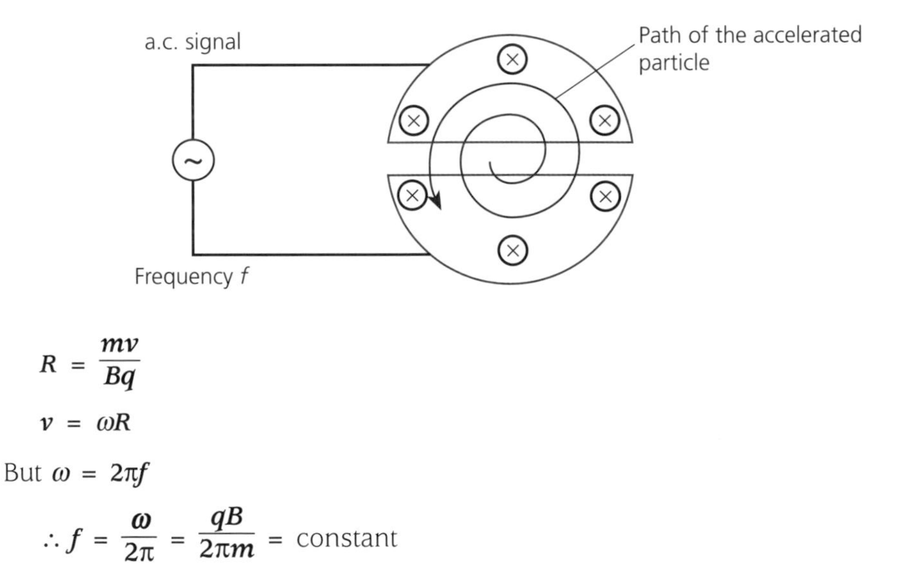

$$R = \frac{mv}{Bq}$$

$$v = \omega R$$

But $\omega = 2\pi f$

$$\therefore f = \frac{\omega}{2\pi} = \frac{qB}{2\pi m} = \text{constant}$$

With a constant frequency a.c. signal across the D-magnets, charged particles accelerate from one D to the next, gain energy each revolution, and move to orbits of larger radii.

The loudspeaker

An a.c. signal on a coil in a radial magnetic field makes the coil move in and out as the current in the coil changes direction. The loudspeaker converts this, changing audio electrical signal into sound waves.

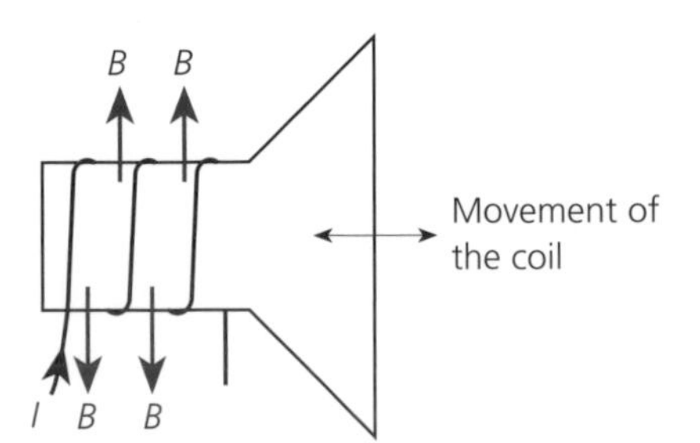

Worked examples

Q1 Electrons enter crossed magnetic and electric fields, and the magnetic field strength is 0.1 T. What magnitude of electric field will select electrons with a velocity of 4.2×10^6 m s^{-1}? What potential must be placed across parallel plates 0.01 m apart to obtain this field?

$v = \frac{E}{B}$ $\quad 4.2 \times 10^6 = \frac{E}{0.1}$ $\quad E = 4.2 \times 10^5$ V m^{-1}

$E = 4.2 \times 10^5 = \frac{V}{d}$ $\quad V = 4.2 \times 10^5 \times 0.01 = 4.2 \times 10^3$ V

Q2 A cyclotron has a maximum radius of 12 cm and a magnetic field of 1.3 T. What is the maximum velocity a proton could reach?

$R = \frac{mv}{Bq}$ $\quad v = \frac{RBq}{m} = \frac{0.12 \times 1.3 \times 1.6 \times 10^{-19}}{1.67 \times 10^{-27}} = 1.5 \times 10^7$ m s^{-1}

Q3 In a simple mass spectrometer, ^{16}O ions with a mass of 16 u travel in a radius of 14.6 cm. What is the mass of O ions that travel with a radius 16.43 cm?

mass = constant $\times R$ $\quad$ 16 u = constant $\times$ 14.6

mass $= \frac{16}{14.6} \times 16.43 = 18$ u

You should now know:

- **how charges are accelerated by electric fields**
- **the definition of the electronvolt**
- **the operation of a velocity selector, mass spectrometer, cyclotron and loudspeaker**

8 Electromagnetic induction

Whenever lines of magnetic flux cut across a wire, forces are generated on the charges in the wire and an e.m.f. is generated which sends a current around the circuit. If the wires do not form a circuit then an e.m.f. is generated across the ends.

Electromagnetic induction effects are only present while changes in flux are taking place.

8.1 Magnetic flux

With a coil of one turn, the flux through the coil is:

flux = $B \times A$

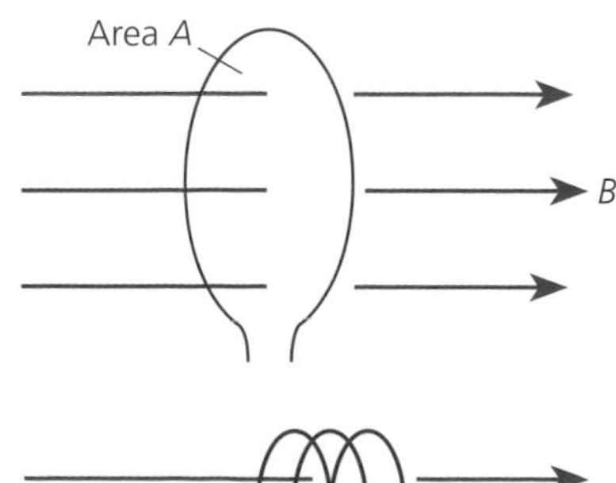

For a coil of N turns, the flux through the coils is:

flux = $B \times A \times N$

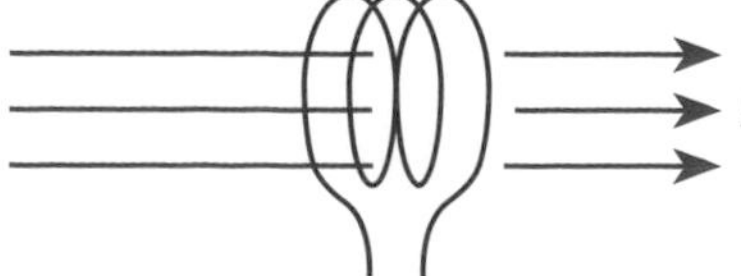

8.2 Faraday's law and Lenz's law

The magnitude of the induced e.m.f. is given by **Faraday's law:**

- **the induced e.m.f. across a conductor is equal to the rate at which magnetic flux is cut by the conductor**

The direction of the induced e.m.f. is given by **Lenz's law:**

- **the direction of any induced e.m.f. is such as to oppose the change that causes it**

The two laws are both covered by the mathematical statement:

$$E = -\frac{d\phi}{dt} \qquad \phi = \text{flux}$$

Methods of changing flux cutting a circuit

There are two methods of making magnetic flux cut a circuit:

1 Keep the circuit stationary and change B.

A simple rule for obtaining the direction of the induced magnetic field in a coil.

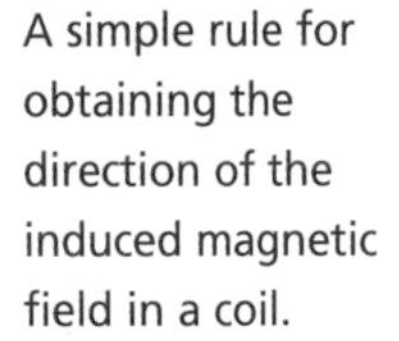

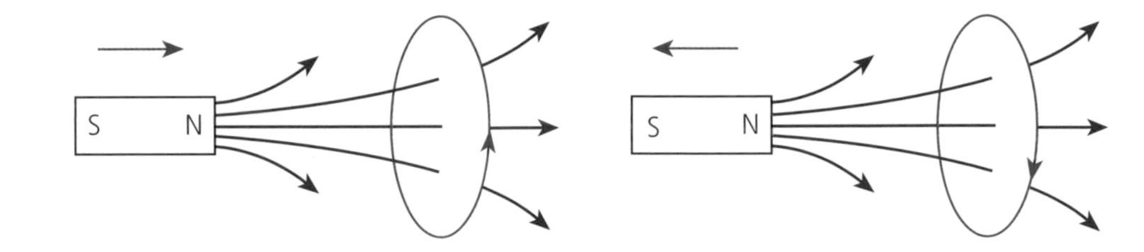

2 Keep B constant and move the circuit.

You should note that in all these systems the resultant e.m.f. generates magnetic fields that oppose the changes producing them.

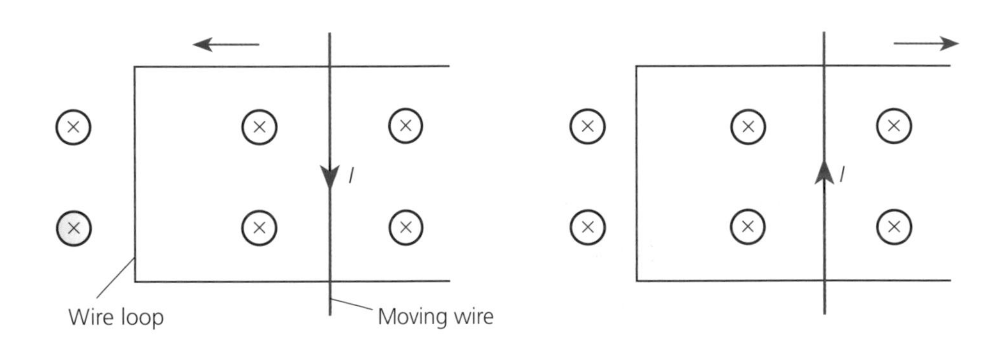

8.3 Self-inductance

When an e.m.f. is applied to a coil of wire, the current through the coil increases, creating a magnetic field. This changing magnetic field creates changing magnetic flux through the coil and generates a self-induced e.m.f. in the coil. This e.m.f. is in such a direction as to oppose the increase in the current.

This self-induced e.m.f. is often called a **back e.m.f.**, since it is in the reverse direction to the e.m.f. generating the original current.

The magnitude of the e.m.f. generated depends on the shape of the solenoid.

e.m.f. $\propto \frac{d\phi}{dt}$, but since $\phi \propto I$, the e.m.f. $\propto \frac{dI}{dt}$ so that:

$$\text{e.m.f.} = L\frac{dI}{dt}$$

where L is a constant called the **self-inductance.** L has units of V s A^{-1} or the henry (H).

1 henry is a large unit, and we often use mH. 1 mH = 1×10^{-3} H.

8.4 Current growth in an inductance

If an inductance is placed in a circuit with a series resistance, when the circuit is connected the current rises in a similar way to the rise in voltage across a capacitor.

$$I = I_0(1 - e^{-\frac{Rt}{L}})$$

The resistance is usually the resistance of the coil of wire forming the inductance.

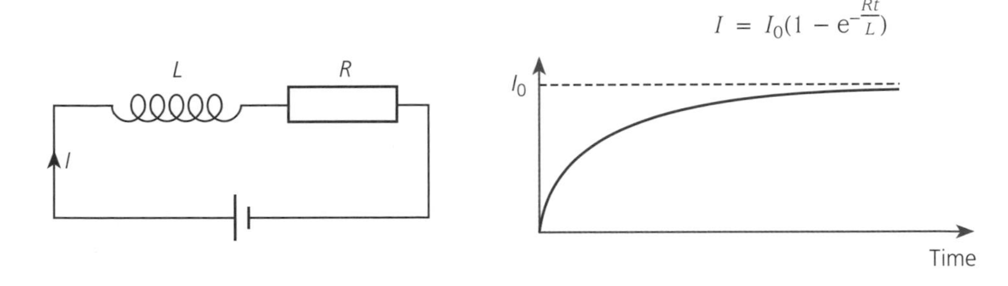

The steady-state condition is reached when the back e.m.f. from the inductor is zero, no flux changing.

If a short is placed across the inductance when the steady current is flowing, the current falls as follows:

$$I = I_0 e^{-\frac{Rt}{L}}$$

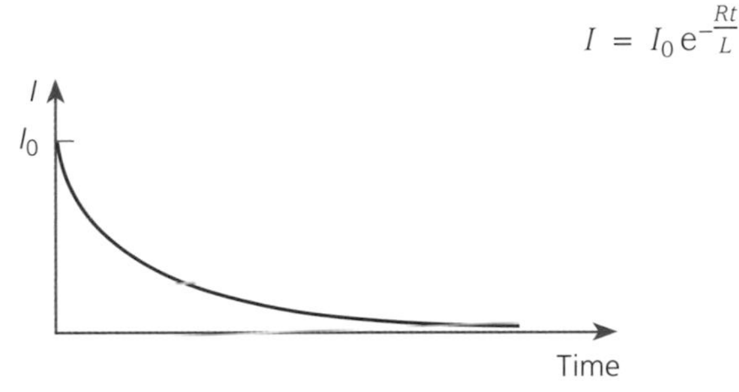

Worked examples

Q1 A car travels along a road at a speed of $55\,\text{m s}^{-1}$ and the Earth's magnetic field has a vertical component of 0.6×10^{-4} T. What is the e.m.f. generated across the axles, separated by a distance of 2.0 m?

$$E = \frac{d\phi}{dt}$$

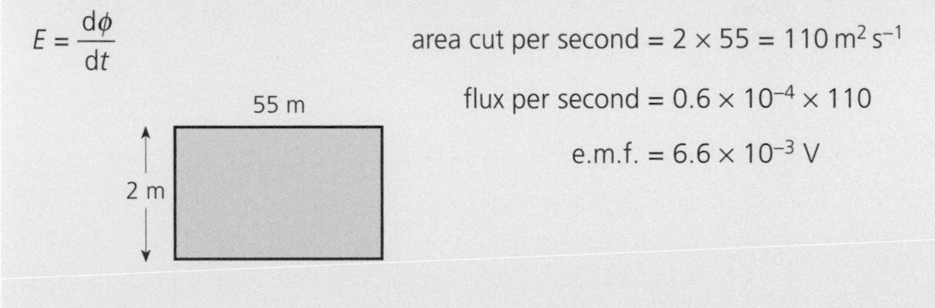

area cut per second = $2 \times 55 = 110\,\text{m}^2\,\text{s}^{-1}$

flux per second = $0.6 \times 10^{-4} \times 110$

e.m.f. = 6.6×10^{-3} V

Q2 Repeat the above calculation for an aircraft of wingspan 50 m travelling at $650\,\text{km h}^{-1}$.

Convert km h^{-1} to m s^{-1}: $\frac{650 \times 10^3}{3600} = 181\,\text{m s}^{-1}$

area cut per second = $181 \times 50 = 9028\,\text{m}^2\,\text{s}^{-1}$

flux per second = $9028 \times 0.6 \times 10^{-4}$ ∴ e.m.f. = 0.54 V

Q3 A coil of inductance 24 mH and resistance 150 Ω has a steady current of 1.4 A flowing through it. If the ends of the coil are shorted together, how long does it take the current to fall to 10% of its initial value?

$$I_0 = 1.4\,\text{A}$$

$$10\%\ I_0 = 0.14\,\text{A}$$

$$I = I_0 e^{-\frac{Rt}{L}}$$

$$\frac{0.14}{1.4} = e^{-\frac{Rt}{L}}$$

$$0.1 = e^{-\frac{150t}{24 \times 10^{-3}}}$$

$$\ln 0.1 = -\frac{150\,t}{24 \times 10^{-3}}$$

$$-2.3 = \frac{150\,t}{24 \times 10^{-3}} -6250\,t$$

$$t = \frac{2.3}{6250} = 3.7 \times 10^{-4}\,\text{s}$$

You should now know:

- the various ways magnetic flux can cut wires in a circuit, and the effect produced
- Faraday's and Lenz's laws
- the concept of self-inductance and how to perform simple calculations
- the way the current changes in a simple *LR* circuit

TOPIC 9

Quantum physics

1 *The photoelectric effect and photons*

Topic 7, Section 9 deals with diffraction and interference when electromagnetic waves, i.e. light, interact with small objects. This is explained in terms of waves and the constructive and destructive interference of the waves after travelling different paths.

However, a number of experiments suggest that electromagnetic waves possess particle properties. The most revealing is the **photoelectric** experiment.

1.1 The photoelectric effect

When a clean metal surface is illuminated with electromagnetic waves of suitable wavelength, electrons (photoelectrons) are released from the surface. When placed in a circuit with a battery, the emitted photoelectrons are accelerated to the positive electrode and a current flows. The magnitude of the current is proportional to the intensity of the illumination.

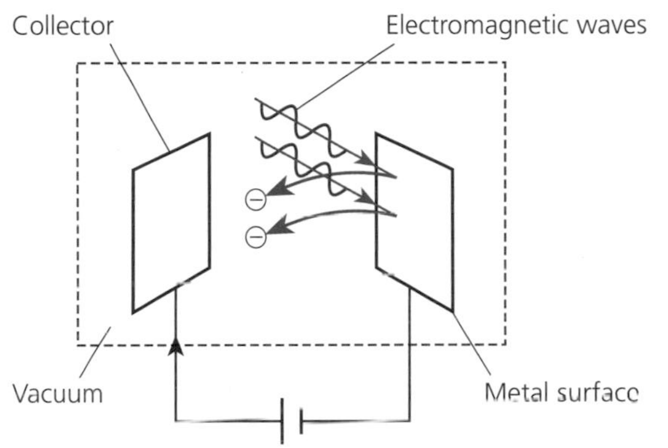

The photoelectric experiment

In order to make some measurements on the relationship between the photoelectrons and the incident illumination, the above apparatus is used with the terminals of the battery reversed, so that the emitted photoelectrons are in an electric field opposing their motion.

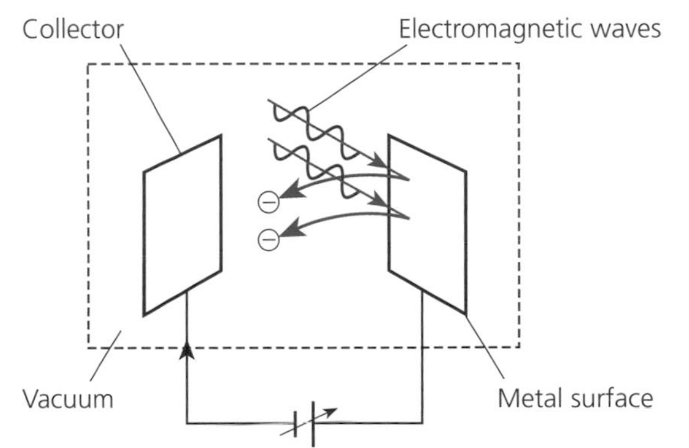

Using this system, the energy of the photoelectrons can be obtained by measuring the potential required to just stop the flow of photoelectrons — the **stopping potential, V_s**.

A photoelectron emitted from the surface with speed v, kinetic energy $\frac{1}{2}mv^2$, will just not reach the collecting electrode when all its kinetic energy has been converted into potential energy.

See Topic 8, Section 7 on the electronvolt to confirm this equation.

$$\text{kinetic energy} = \tfrac{1}{2}mv^2 = eV_s$$

Results of the experiment

The experiment was performed with electromagnetic waves of different wavelengths and intensities, and it was expected that the electrons would be always emitted with a spread of energies.

The following unexpected results were obtained:

- **The number of photoelectrons emitted depends on the intensity of the illumination, provided the frequency of the electromagnetic radiation is above a certain threshold value. If the frequency is below this value, no photoelectrons are emitted no matter how intense the illumination.**
- **The maximum kinetic energy of the emitted photoelectrons is independent of the intensity of the illumination but proportional to the frequency, provided it is above the threshold value.**
- **When different metals are illuminated, the threshold frequency is different.**

The results are summarised in the following graph.

Remember, $c = f\lambda$. Increasing frequency means decreasing wavelength.

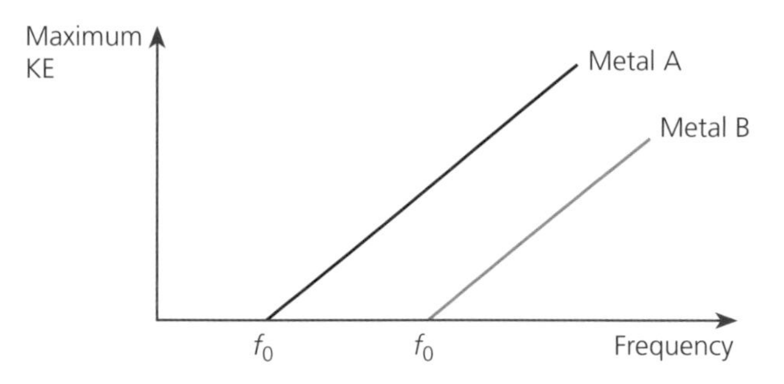

Metals such as sodium and caesium have threshold frequencies in the visible part of the spectrum, whereas zinc requires frequencies in the ultraviolet part of the spectrum.

1.2 Quantum theory

See Topic 4, Section 8.3.

When objects are heated, they emit electromagnetic radiation. At low temperatures, they are a dull red and change from yellow to white as the temperature increases. In order to explain this, Max Planck introduced the idea of quantum physics in which energy is always emitted or absorbed in discrete units, called **quanta**. Thus, the electromagnetic radiation emitted by the heated body must also be in discrete units of energy, called **photons**.

The energy of each quantum and hence each photon is $E = hf$, where h is the Planck constant ($h = 6.63 \times 10^{-34}$ J s).

1.3 Einstein photoelectric equation

This concept was used by Einstein to explain the results of the photoelectric experiment. A unit of electromagnetic energy is called a **photon**.

The incident photon strikes the metal with a certain energy, $E = hf$.

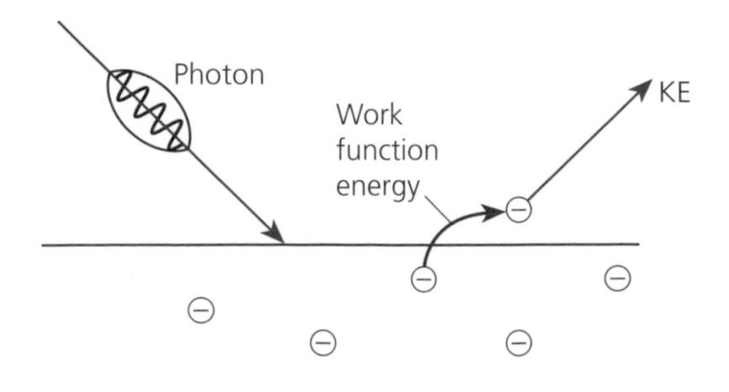

This energy is given to one electron in the metal. Some of the energy is used to release the electron from the surface — the **work function energy**. Any remaining energy appears as kinetic energy of the emitted photoelectron.

The energies involved are very small, so it is often more convenient to work in eV. Work functions are nearly always quoted in eV.

incident photon energy = work function energy + maximum KE of photoelectron

$hf = \phi + \frac{1}{2}mv^2_{max}$

At the threshold frequency, KE = 0 $\therefore hf_0 = \phi$

$hf = hf_0 + \frac{1}{2}mv^2_{max}$ **This is the Einstein photoelectric equation.**

Worked examples

Q1 Calculate the energy in eV of photons of wavelength 550 nm.

$c = f\lambda \qquad f = \frac{c}{\lambda} = \frac{3 \times 10^8}{550 \times 10^{-9}} = 5.45 \times 10^{14}\ \text{Hz}$

$E = hf = 6.63 \times 10^{-34} \times 5.45 \times 10^{14} = 3.6 \times 10^{-19}\ \text{J}$

$E = \frac{3.6 \times 10^{-19}}{1.6 \times 10^{-19}} = 2.3\ \text{eV}$

Q2 How many photons are emitted per second from a monochromatic ultraviolet lamp of wavelength 200 nm when emitting 100 W of ultraviolet radiation?

$E = \frac{hc}{\lambda} = \frac{6.63 \times 10^{-34} \times 3 \times 10^8}{200 \times 10^{-9}} = 9.95 \times 10^{-19}\ \text{J}$

$100\ \text{W} = 100\ \text{J s}^{-1} = \frac{100}{9.95 \times 10^{-19}} = 1.0 \times 10^{20}$ photons

Q3 A metal surface has a work function of 4.7 eV. It is illuminated by photons of energy 5.6 eV. What is the energy of the emitted photoelectrons in joules and the wavelength of the incident illumination?

$hf = \phi + \frac{1}{2}mv^2_{max}$

$5.6 = 4.7 + \frac{1}{2}mv^2_{max} \qquad \therefore \frac{1}{2}mv^2_{max} = 5.6 - 4.7 = 0.9\ \text{eV}$

$= 0.9 \times 1.6 \times 10^{-19} = 1.4 \times 10^{-19}\ \text{J}$

$E = \frac{hc}{\lambda} \quad \lambda = \frac{hc}{E} = \frac{6.63 \times 10^{-34} \times 3 \times 10^8}{5.6 \times 1.6 \times 10^{-19}} = 2.22 \times 10^{-7}\ \text{m} = 220\ \text{nm}$

You should now know:

- **the electrical circuit for a photocell and the modification for the photoelectric experiment**
- **what is measured in the photoelectric experiment and the results**
- **the explanation of these results using quantum physics**
- **the Einstein equation**

2 *Wave–particle duality*

A further experiment that suggests electromagnetic waves possess particle properties is the Compton effect. In this experiment, electromagnetic photons collide with stationary electrons, which then move off, having gained energy as a result of the collision.

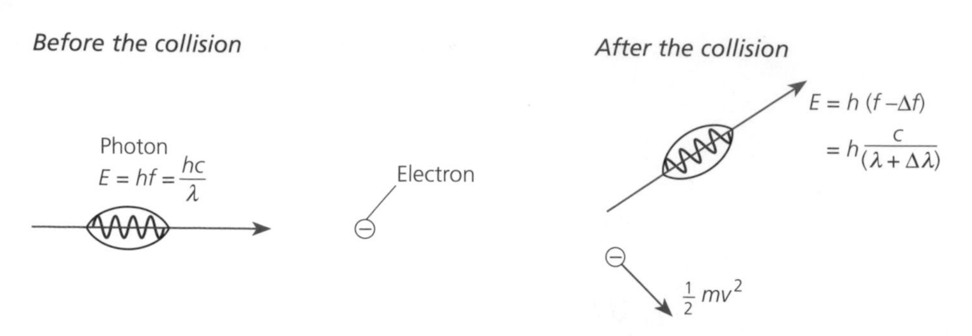

The collisions are found to obey the laws of conservation of energy and momentum as in particle collisions (see Topic 4, Section 3). So after the collision, if the photon has transferred some energy/momentum to the electron, the energy/momentum of the photon is reduced.

Since the energy of the photon is $E = hf$, when E reduces, so must f, and since velocity $= f\lambda$ and the velocity is constant, the wavelength of the scattered photon will be greater.

When λ of the scattered photons was measured, this was found to be the case, so this experiment also suggests that electromagnetic waves possess particle properties.

2.1 Wave properties of particles

Since light shows wave and particle properties, this suggests that particles may possess wave properties.

Electron diffraction

When electrons of suitable energy are fired at thin metal foils, the electrons are scattered. The spatial distribution of scattered electrons is not random, but forms a series of concentric rings.

Other particles – protons, neutrons etc – exhibit diffraction when scattered by suitable systems.

A plot of the density distribution of electrons on the screen is similar to that obtained when light waves are diffracted by a circular aperture.

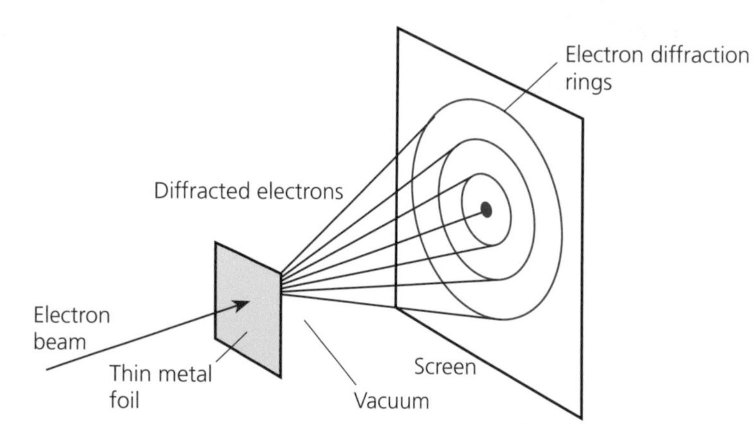

This suggests that the electrons possess wave properties, and what is observed in the experiment is **electron diffraction**.

Electron diffraction and probability

If the beam of electrons is reduced in intensity, the pattern of electrons on the screen appears random and, for each electron, all we can state is the probability that an electron will land at a particular point. The intensity of the diffraction pattern is highest where the probability is highest.

2.2 Wave–particle duality – the de Broglie equation

Clearly a link is required between these two descriptions, and this is provided by the de Broglie equation.

Considering photons, de Broglie was able to show that the link between the wave description of light and the particle description required by the photoelectric and Compton effect experiments is as follows:

$$\lambda = \frac{h}{p}$$

where λ is the wavelength of the photon, and p is its momentum considering it as a particle.

De Broglie suggested that this equation should also apply to particles. The de Broglie equation is:

$$p = \frac{h}{\lambda}$$

where p is the momentum of the particle, and λ is its wavelength considering it as a wave and is termed the **de Broglie wavelength**.

Note that Planck's constant is the link between the two pictures, wave and particle.

Question Are electrons waves or particles? Is light waves or particles?

Answer It depends on the experiment being considered: in some experiments, a wave picture must be used; in some experiments, a particle picture is required.

The relationship between energy and wavelength

Particles
High-energy particles have a high momentum and short wavelengths.
Low-energy particles have a low momentum and long wavelengths.

Electromagnetic waves
Short-wavelength photons have a large momentum and large kinetic energy.
Long-wavelength photons have a small momentum and small kinetic energy.

Worked examples

Q1 Calculate the momentum of visible photons with a wavelength of 550 nm, and of X-ray photons of wavelength 1 nm.

$$p = \frac{h}{\lambda} = \frac{6.63 \times 10^{-34}}{550 \times 10^{-9}} = 1.2 \times 10^{-27}\,\text{kg m s}^{-1}$$

$$p = \frac{h}{\lambda} = \frac{6.63 \times 10^{-34}}{1 \times 10^{-9}} = 6.63 \times 10^{-25}\,\text{kg m s}^{-1}$$

Q2 A photon beam of wavelength 650 nm strikes a screen. If the beam has a flux of 1×10^{8} photons per second, calculate the force on the screen.

momentum of each photon $p = \frac{h}{\lambda} = \frac{6.63 \times 10^{-34}}{650 \times 10^{-9}} = 1.02 \times 10^{-27}$ kg m s^{-1}

force = rate of change of momentum = no. of photons per sec and × momentum

$= 1 \times 10^{8} \times 1.02 \times 10^{-27} = 1.0 \times 10^{-19}$ N

Q3 An electron has an energy of 500 keV. Calculate the energy of the electron in joules, its momentum and associated wavelength.

energy in joules = $500 \times 10^3 \times 1.6 \times 10^{-19} = 8.0 \times 10^{-14}$ J

$KE = \frac{1}{2}mv^2 = \frac{p^2}{2m}$

$p = \sqrt{2mKE}$

$p = \sqrt{2 \times 9.11 \times 10^{-31} \times 8.0 \times 10^{-14}}$

$p = 3.82 \times 10^{-22}$ kg m s^{-1}

$\lambda = \frac{h}{p} = \frac{6.63 \times 10^{-34}}{3.82 \times 10^{-22}} = 1.7 \times 10^{-12}$ m

You should now know:

- **an outline of the Compton effect experiment**
- **the set-up for electron diffraction and what is observed**
- **the de Broglie equation and its use in wave particle duality**

3 *Electromagnetic waves*

Electromagnetic waves are oscillations of interacting electric and magnetic fields.

Electromagnetic waves are unlike other waves in that they will travel through a vacuum as well as through a medium. Visible light is an electromagnetic wave.

The oscillations of the electric and magnetic fields in an electromagnetic wave are at 90° to each other.

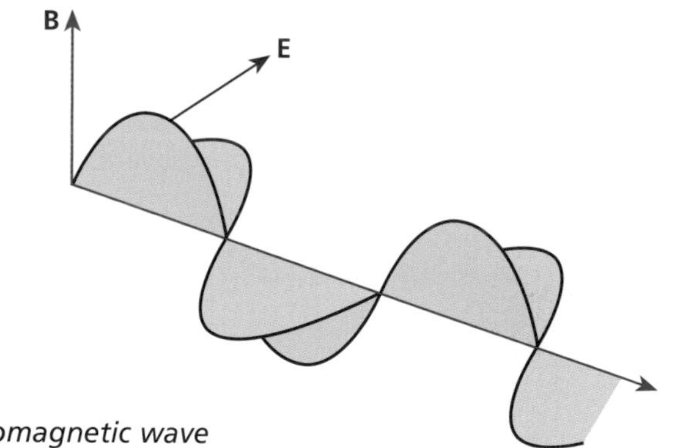

A polarised electromagnetic wave

Electromagnetic waves can travel through a vacuum because a changing electric field generates a magnetic field and a changing magnetic field generates an electric field, so energy is transferred back and forth between the two fields.

3.1 The velocity of electromagnetic waves in a vacuum

The velocity of electromagnetic waves in a vacuum is a constant and, as the velocity is very large, its value depends on our definition of the standard metre. Nowadays, we use the speed of light as one definition of the standard metre by stating that:

1 metre is the distance travelled by an electromagnetic wave in a vacuum in $\frac{1}{299\,792\,458}$ s.

Hence the velocity of electromagnetic waves in a vacuum, c = 299 792 458 m s^{-1}.

When performing calculations using the velocity of light, a value of 3.0×10^8 is acceptable.

Velocity of electromagnetic waves in a medium

As we have seen in Topic 7, Section 7, the velocity of light in a medium decreases, and the refractive index of the medium is a measure of the reduction.

$$\text{refractive index, } n = \frac{\text{the velocity of light in a vacuum}}{\text{the velocity of light in the medium}}$$

3.2 The electromagnetic spectrum

Electromagnetic waves obey the same equation as all waves:

velocity = frequency × wavelength ($c = f\lambda$)

Electromagnetic waves cover a large range of wavelengths and hence frequencies, and the complete range of waves that can be produced is called the **electromagnetic spectrum**.

The spectrum is so large that electromagnetic waves are initially grouped in broad ranges depending on (a) how they are produced, or (b) how they are detected.

We can use either wavelength or frequency to specify the part of the electromagnetic spectrum being considered. In the visible part of the spectrum, it is common practice to use the wavelength, but we could equally well use frequency if required.

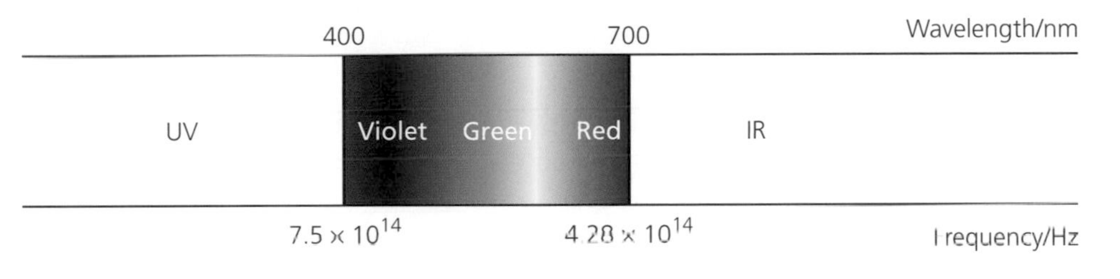

The complete electromagnetic spectrum

Below is a simple subdivision of the electromagnetic spectrum. Note that some regions of the subdivisions overlap, e.g. γ-rays and X-rays. The only difference between these two is the method of their production.

Many of these subdivisions are broken down into smaller divisions. For example, radio waves are further subdivided into VHF, UHF, long, medium and short radio waves.

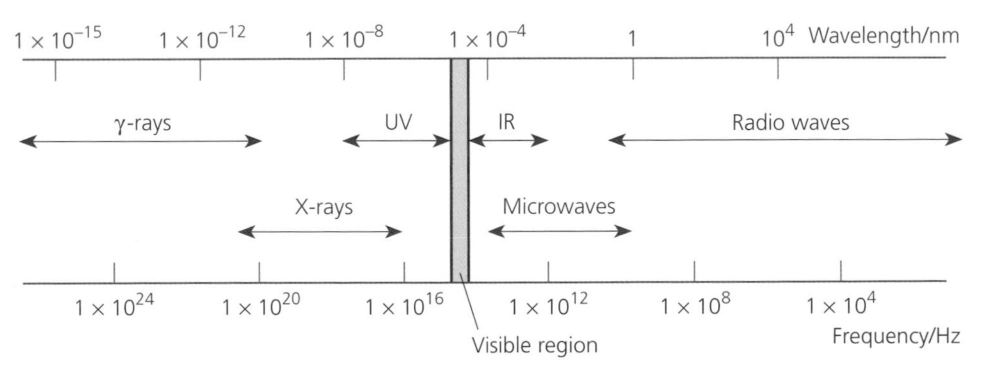

The production and detection methods in some cases are only examples and are not comprehensive.

- **γ-rays** are produced when atomic nuclei need to lose energy. They are detected by the ionisation they produce in particle counters, such as Geiger–Müller tubes, in scintillation counters, or the blackening of photographic film.
- **X-rays** are produced when high-energy electrons (200 keV) collide with heavy metal targets. Electrons orbiting the atoms are knocked out and the subsequent rearrangement of the remaining electrons in the energy levels emits photons of electromagnetic radiation. They are detected by scintillation counters or the blackening of photographic film.

- **Ultraviolet, visible and infrared** radiation is produced by electronic transitions in atoms, molecules or semiconductors when heated or suitably provided with energy. Depending on the actual type of radiation, this can be detected by the eye, photographic film or heat detectors.
- **Microwaves** are produced by special electronic devices called Klystrons or the vibrations of molecules. They are detected by observing the oscillation of vibrating molecules.
- **Radio waves** are produced by the oscillation of electrons in electric fields in suitable wires, e.g. aerials. They are detected by the oscillations generated in suitable tuned electrical circuits.

Worked examples

Q1 γ-rays have wavelengths in the range 1.0×10^{-16} to 1.0×10^{-11} m. What is the highest energy photon γ-ray?

Since $E = hf = \frac{hc}{\lambda}$ the highest energy has the shortest wavelength.

$$E = \frac{6.63 \times 10^{-34} \times 3.0 \times 10^{8}}{1.0 \times 10^{-16}}$$

$$E = 2.0 \times 10^{-9}\ \text{J}$$

Remember, for diffraction to be observed, the diffracting object must be of a similar size to the wavelength (see Topic 7, Section 9.1).

Q2 X-rays have a frequency of 1.0×10^{17} Hz. What is their wavelength? If you wanted to observe X-ray diffraction, what would form a suitable grating for these X-rays?

$c = f\lambda \qquad 3.0 \times 10^{8} = 1.0 \times 10^{17} \times \lambda \qquad \lambda = 3.0 \times 10^{-9}$ m

Atoms in crystals have a similar spacing to this and so could be used as 3D diffraction gratings.

Q3 In certain parts of the country, Radio 4 is transmitted at a frequency of 96.0 MHz. What is the wavelength of these radio waves?

$c = f\lambda \qquad 3.0 \times 10^{8} = 96 \times 10^{6} \times \lambda \qquad \lambda = 3.1$ m

You should now know:

- **the form and the velocity of electromagnetic waves**
- **the form of the electromagnetic spectrum and the names of the various regions**
- **the explanation for the origin of the waves in the various regions; the detection methods used**

4 *Spectra*

Spectra can also be obtained by heating the gas in a flame. The characteristic spectra of solids and liquids can also be obtained by heating them so that they vaporise.

When a large voltage is placed across two electrodes in a tube containing a gas at low pressure, i.e. a discharge tube, electromagnetic radiation is released.

The colour seen is characteristic of the atoms of the gas in the discharge tube.

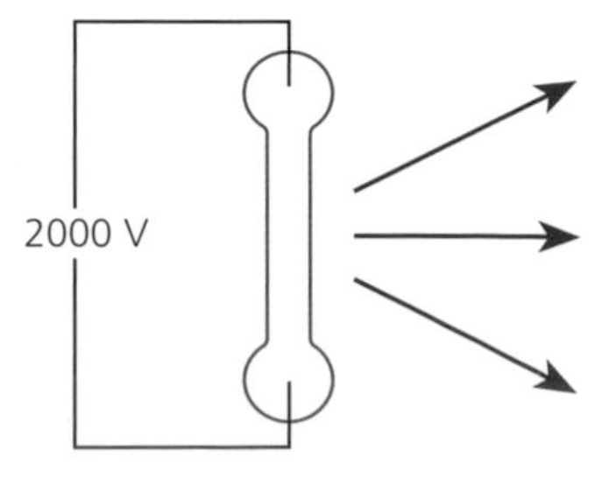

If the radiation emitted is observed with a spectrometer, it is found that the spectrum consists of series of lines of different colours, i.e. different wavelengths.

4.1 The hydrogen spectrum

The simplest atom is a hydrogen atom, which consists of a positive core, the nucleus, around which a single electron orbits.

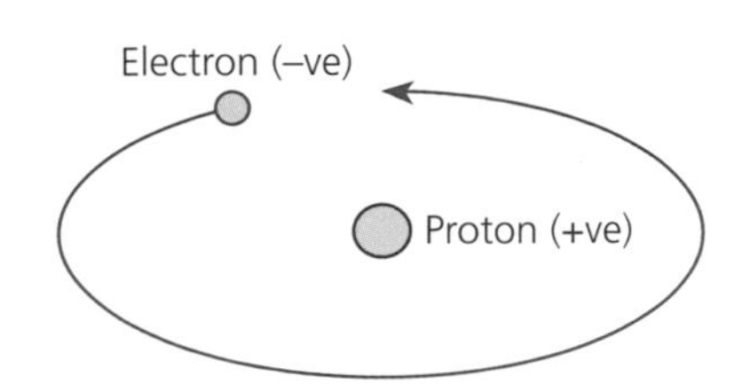

Hydrogen produces the simplest spectrum, consisting of a series of spectral lines with similar characteristics.

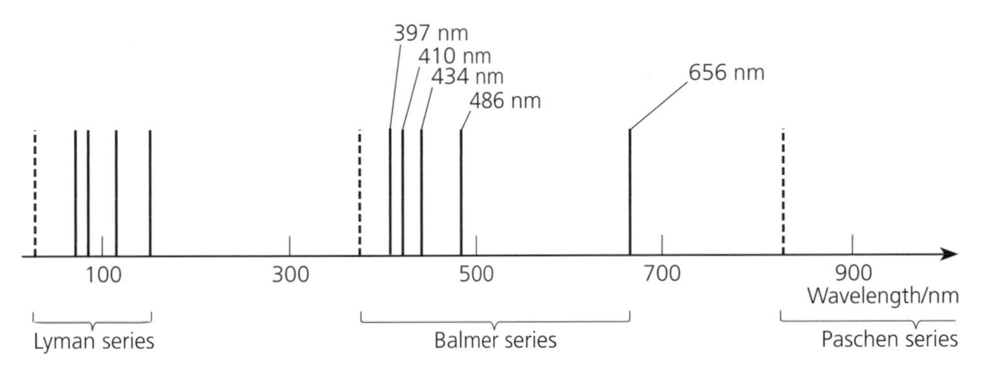

Sodium street lamps look yellow owing to two intense yellow lines found in its spectrum.

One series, the Balmer series, has four lines in the visible part of the electromagnetic spectrum, consisting of an intense red line, a weaker blue/green line and two weak violet lines. The visible light observed from hydrogen excited in a discharge tube is a mixture of these colours.

Origin of the hydrogen spectrum

Allowed orbits have circumferences equal to a whole number of electron waves.

When an electron orbits the hydrogen nucleus, only certain orbits are allowed with fixed radii.

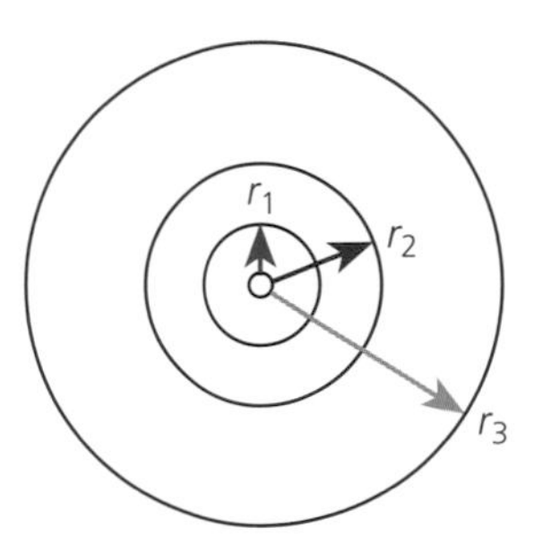

In each of these orbits, the electron has a certain total energy: kinetic energy, plus electrostatic potential energy. To move an electron from an orbit of small radius to an orbit of larger radius, energy must be supplied. When the reverse process takes place, energy is released.

4.2 Energy-level diagrams

The changes in energy of the electron are usually represented by energy-level diagrams. Electrons in hydrogen sit at the lowest energy level. When electrons are excited to an upper energy level, after a short time they fall down to a lower level, releasing energy in the form of photons of electromagnetic radiation.

If photons of the correct wavelength are incident on an atom, they can be absorbed by the atom and excite electrons to higher energy levels.

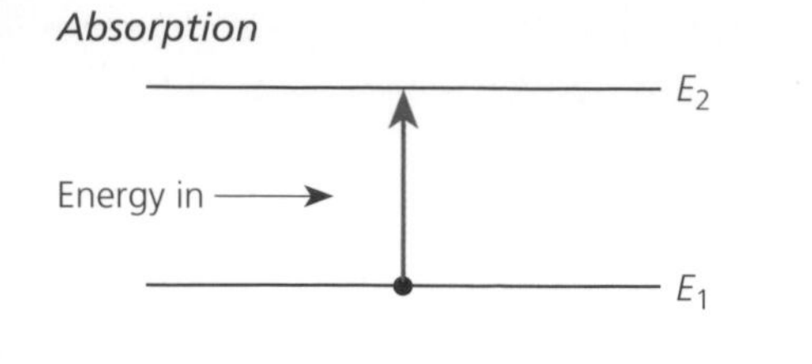

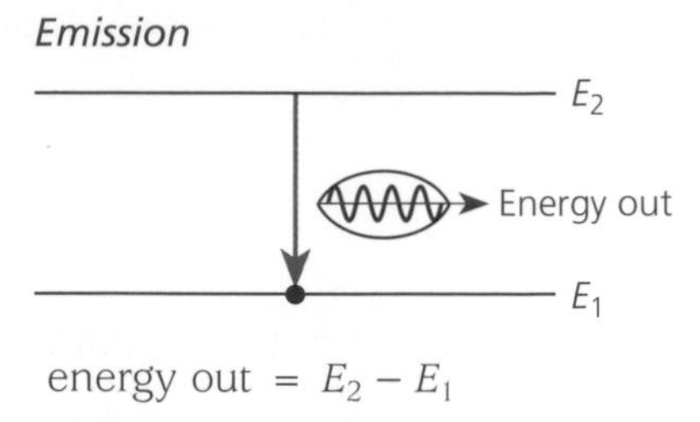

energy out $= E_2 - E_1$

This energy is converted to a photon: $E_2 - E_1 = hf$

$$E_2 - E_1 = \frac{hc}{\lambda}$$ λ = wavelength of the emitted photon

There are an infinite number of energy levels between the ground state and the energy of an ionised atom.

Ground state and the ionisation energy

The lowest allowed energy state in hydrogen is called the ground state and is the energy level normally occupied by the electron. As energy is supplied to the electron, it moves to new orbits of higher energy and larger radii. The energies of these larger radii orbits form a series of energy levels.

When an electron is removed completely from the atom, excited from energy level E_1 to E_∞, the atom is said to be ionised. The energy required is called the **ionisation energy**.

In the energy-level diagram, the zero of energy is that of an ionised atom.

In the case of hydrogen, the ionisation energy is 13.6 eV.

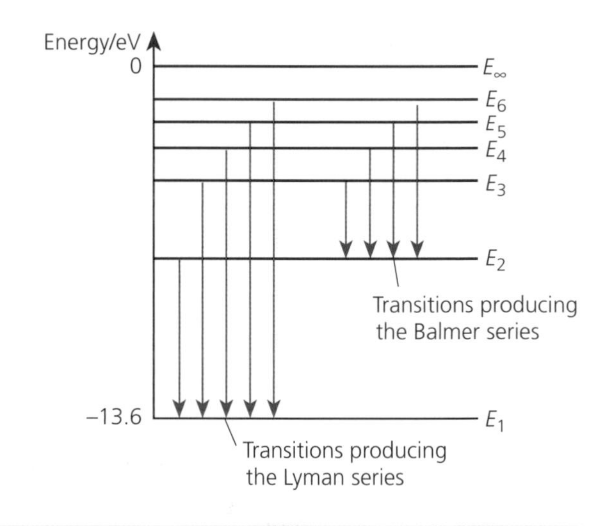

Worked examples

Q1 The red line in the hydrogen spectrum has a wavelength of 656 nm. What is the energy difference between the two energy levels that generate this spectral line?

$$\Delta E = \frac{hc}{\lambda} = \frac{6.63 \times 10^{-34} \times 3.0 \times 10^{8}}{656 \times 10^{-9}} = 3.03 \times 10^{-19}\text{ J} = 1.90\text{ eV}$$

Q2 The yellow radiation from a sodium lamp consists of two lines of wavelengths 589.0 nm and 589.6 nm. Calculate the energy difference for each of the transitions.

$$\Delta E = \frac{6.626 \times 10^{-34} \times 2.997 \times 10^{8}}{589.0 \times 10^{-9}} = 3.371 \times 10^{-19}\text{ J}$$

$$\Delta E = \frac{6.626 \times 10^{-34} \times 2.997 \times 10^{8}}{589.6 \times 10^{-9}} = 3.368 \times 10^{-19}\text{ J}$$

difference $= 0.003 \times 10^{-19}$ J

Q3 What wavelength photon must be incident on unexcited hydrogen to produce an ionised atom?

$$E_{\infty} - E_1 = 13.6 \text{ eV} = 13.6 \times 1.60 \times 10^{-19} = 2.18 \times 10^{-18} \text{ J}$$

$$2.18 \times 10^{-18} \text{ J} = \frac{6.63 \times 10^{-34} \times 3.00 \times 10^{8}}{\lambda}$$

$$\lambda = 9.12 \times 10^{-8} \text{ m} = 91.2 \text{ nm}$$

Always remember to convert from electronvolts to joules when performing these calculations.

You should now know:

- **how atomic spectra are produced**
- **the general form of atomic spectra**
- **the form of the hydrogen spectrum**
- **energy-level diagrams and how to calculate the wavelength emitted**
- **the energy-level diagram for hydrogen and ground state and ionisation energy**

5 *Probing matter—scattering experiments*

Matter is made up of atoms and molecules. A whole range of diffraction/scattering experiments are performed in order to measure the following:

- the distribution of atoms in a solid
- the structure of atoms
- the distribution of electrons in an atom
- the shape, size and charge distribution of the nucleus

The words diffraction and scattering are often interchanged in this area, but all the experiments follow a similar pattern.

The basic rule for all diffraction/scattering experiments is that the wavelength of the wave or the de Broglie wavelength of the particle must be of the same order as the spacing/size of the diffracting/scattering object being observed.

5.1 De Broglie wavelength of particles

Kinetic energy = $\frac{1}{2}mv^2$, hence momentum of the particle $p = mv = \sqrt{2 \times \text{KE} \times m}$.

The de Broglie equation states $p = \frac{h}{\lambda}$ so $mv = \frac{h}{\lambda} = \sqrt{2\text{KE}m}$.

The *larger* the kinetic energy of the particle, the *smaller* the de Broglie wavelength.

5.2 Geiger–Marsden experiment—atomic structure

Prior to performing this experiment, the atom was thought to be a mixture of +ve and –ve charges. In this experiment, high-energy α particles are fired at a very thin gold film.

An α particle is the nucleus of a helium atom and consists of two protons and two neutrons. It has a charge +2e.

Using a suitable detector (a Geiger Müller counter) the rate at which α particles are detected is measured as a function of the scattering angle θ.

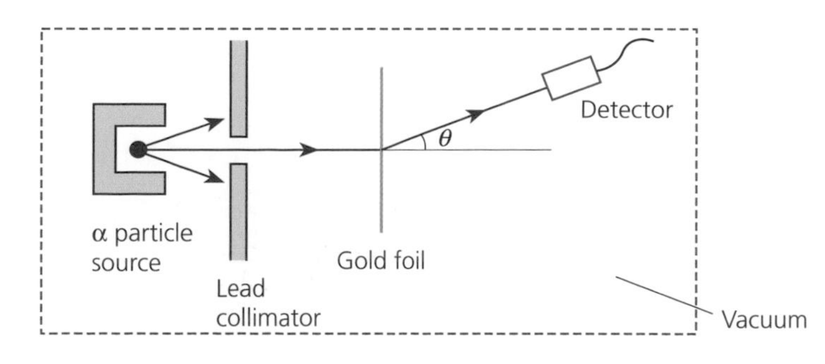

Further experiments showed that the charge at the centre was +ve.

The results of the experiment are as follows:

- The majority of α particles pass through the foil undeflected. **This suggests that much of the foil, and hence the atoms, consists of empty space.**
- A few α particles ae deflected through angles of $>90°$. **This suggests that all the charge of one sign must be concentrated at one place, which must also contain the major portion of the mass of the atom.**

5.3 The Rutherford atom

As a result of these experiments, Rutherford suggested that the atom consists of a positive central core — the nucleus — around which the negative electrons orbit.

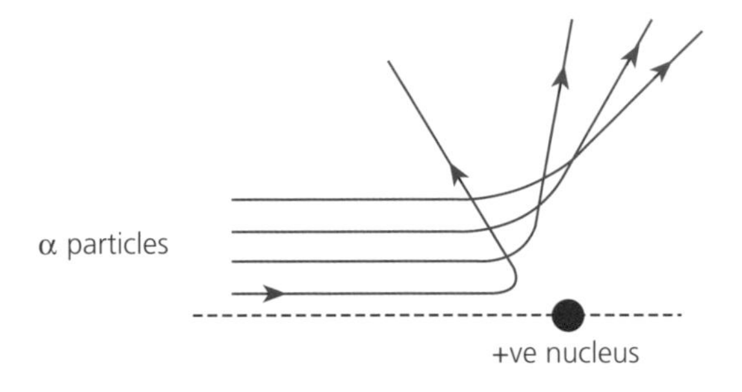

The size of the nucleus is thus thought to be about 10^{-14} m.

The number of the α particles scattered at different angles can be calculated assuming a coulombic repulsion force between the α particles and the nucleus. If the closest distance of approach to the nucleus was less than 10^{-14} m, the coulombic calculation does not give the correct result, suggesting that the α particles are interacting with the nuclear surface.

5.4 X-ray diffraction

X-rays have wavelengths in the region of 10^{-8} to 10^{-11} m. They can be used to measure the distribution of atoms in solids, which are arranged in planes with a spacing of about 10^{-10} m. X-ray electromagnetic waves are reflected off these planes. Constructive interference takes place when the path difference is a whole number of wavelengths. This produces X-ray diffraction spots or lines, depending on the experimental geometry.

X-rays interact with the electrons in atoms, so the more electrons in an atom, the better are the diffraction results.

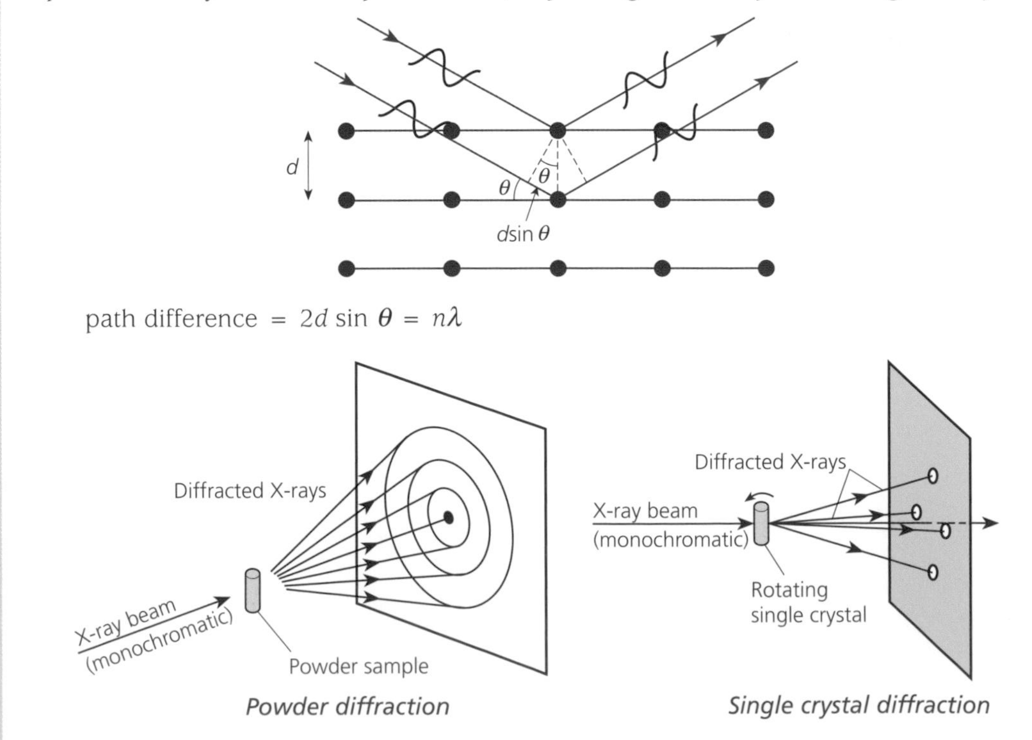

Powder diffraction

Single crystal diffraction

5.5 Neutron diffraction

In nuclear reactors, thermal neutrons with energies of about 0.025 eV have de Broglie wavelengths of 10^{-10} m. They are ideal for diffraction by crystal planes. As they are uncharged, they interact only with the nuclei of atoms and provide better diffraction results for lighter atoms than X-rays.

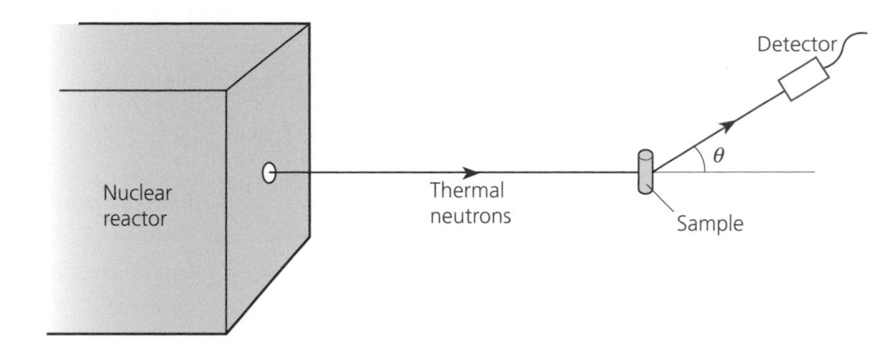

5.6 High-energy electron diffraction

Electrons with energies in the range 10 GeV have de Broglie wavelengths similar to the size of the nucleus. Since they are charged, in scattering experiments they provide information about the charge distribution in the nucleus.

nm on both sides of the equation.

Worked examples

Q1 The spacing of planes in a nickel crystal is 0.17 nm. What is the diffraction angle for X-rays of wavelength 0.10 nm when n = 1?

Using $n\lambda = 2d \sin \theta$.

$$0.10 = 2 \times 0.17 \times \sin \theta$$

$$\sin \theta = \frac{0.10}{2 \times 0.17} \qquad \theta = 17°$$

Q2 Calculate the repulsive force when an a particle is at a distance of 1.0×10^{-14} m from a gold nucleus with a charge of 79e.

2e 79e

1×10^{-14}

$$F = \frac{Q_1 Q_2}{4\pi\varepsilon_0 d^2}$$

$$F = \frac{2 \times 79 \times (1.6 \times 10^{-19})^2}{4 \times \pi \times 8.85 \times 10^{-12}\,(1.0 \times 10^{-14})^2} = 360\ \text{N}$$

Q3 Calculate the de Broglie wavelength of thermal neutrons with an energy 0.025 eV.

$$0.025\ \text{eV} = 0.025 \times 1.6 \times 10^{-19} = 4.0 \times 10^{-21}\ \text{J}$$

$$\lambda = \frac{h}{\sqrt{2KEm}} = \frac{6.6 \times 10^{-34}}{\sqrt{2 \times 4.0 \times 10^{-21} \times 1.67 \times 10^{-27}}} = 1.8 \times 10^{-10}\ \text{m}$$

You should now know:

- **the information obtained from scattering and diffraction experiments**
- **the α-particle scattering experiment and the results obtained**
- **the form of the Rutherford atom and the size of the nucleus**
- **the neutron diffraction and high-energy electron scattering experiments**

6 The nuclear atom

As a result of scattering experiments, a model for the atom was obtained that consisted of a central positive core — the nucleus — around which electrons orbit.

The nucleus contained two particles or **nucleons**:

1. **A proton carrying a positive charge of 1.6×10^{-19} C (the same as the charge on an electron) and mass 1.673×10^{-27} kg.**
2. **A neutron with no charge and mass 1.675×10^{-27} kg.**

The particles in the nucleus are held together by nuclear forces, which have the following properties:

- **they are short range and are only observed when the distance between protons and neutrons is of the order of 1×10^{-14} m**
- **they are strong forces, so when two protons are closer than 1×10^{-14} m they are stronger than the repulsive coulomb force**
- **they are charge independent, so exist between two protons, two neutrons or a proton and a neutron**

6.1 Neutral atoms

In neutral atoms, the number of positive charges (protons) in the nucleus is equal to the number of electrons orbiting. Hence, the overall charge of such atoms is zero.

The chemical property of an atom is dictated by the number of orbiting electrons and hence the number of protons in the nucleus. This gives rise to the periodic table of the elements, with the elements listed in order of increasing electron and hence proton number.

Atomic number, Z

This is the number of protons in the nucleus of the atom and hence the number of orbiting electrons. The atomic number of hydrogen is 1, sodium 11, uranium 92 etc.

Mass number or nucleon number, A

The mass number of an atom is the total number of protons plus neutrons (nucleons) in the nucleus. The mass number along with the atomic number is used to specify the structure of atoms as follows:

$^{A}_{Z}X$, where X is any element, e.g. $^{12}_{6}C$, $^{23}_{11}Na$, $^{238}_{92}U$

Atomic mass unit

Since the mass of a proton and a neutron are almost the same, a new unit was introduced — the atomic mass unit:

1 u is one-twelfth the mass of the isotope of carbon containing six protons and six neutrons.

$$1\,u = 1.660566 \times 10^{-27}\text{ kg}$$

The mass of a proton is 1.007276 u, a neutron 1.008665 u and an electron 0.000549 u.

6.2 Ionised atoms

In normal circumstances, an ionised atom has lost one or more electrons, so that it has a net positive charge, **a positive ion**.

Under special circumstances, it is possible to add electrons, giving the atom a net negative charge.

6.3 Isotopes

The number of neutrons in the nucleus is not fixed but depends on the atom concerned. Atoms with the same number of protons in the nucleus (the same element) but with different numbers of neutrons are called **isotopes**. The nucleus as a group of protons and neutrons is also called a **nuclide**.

Naturally occurring stable isotopes

There are two forces in a nucleus: a repulsive long-range coulomb force and a short-range attractive nuclear force. Moving up the periodic table, the repulsive force increases considerably as more protons are added to the nucleus, and so in stable nuclei, extra electrons are required to counteract this increasing force.

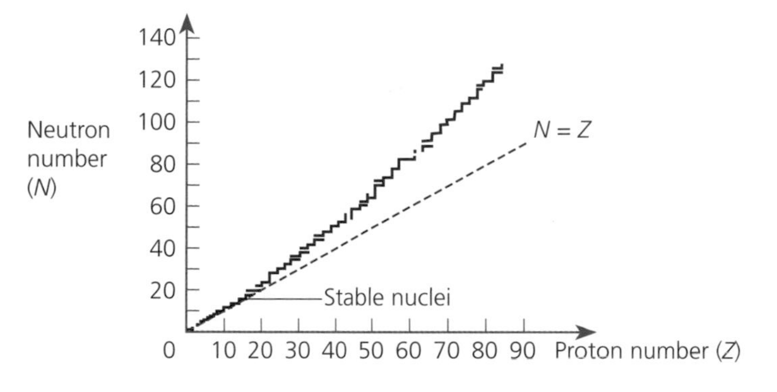

Gold has only one isotope, tin has ten isotopes. Bismuth is the heaviest stable atom in the periodic table.

6.4 Radioactivity

For more information, see Section 7 on radioactivity.

When a nucleus does not have the correct number of protons and neutrons for a stable configuration, or it has too much energy, it is radioactive. The nucleus attempts to reach a stable configuration by the emission of suitable particles or releasing the energy as electromagnetic radiation.

Radioactive isotopes

Natural radioactive isotopes. Many elements contain naturally occurring radioactive isotopes in very small quantities. However, elements in the periodic table above bismuth found naturally are all radioactive.

Artificial radioactive isotopes. It is possible to add neutrons to any nucleus by placing naturally occurring elements in a nuclear reactor, where they are bombarded by a flux of thermal neutrons. Some of the neutrons are captured by nuclei, increasing their neutron number above the figure for a stable nucleus and hence making the material radioactive.

You should now know:

- **the particles in the nucleus and the properties of the strong nuclear force**
- **the nature of ionised atoms**
- **the definition of atomic number, mass number and the term isotope**
- **the definition of atomic mass unit**
- **the shape of the stable isotope curve and why elements are radioactive**

7 *Radioactivity*

Whenever a nucleus has the wrong ratio of neutrons to protons for stability, the nucleus is radioactive and attempts to reach a stable configuration by the emission of particles.

Too many neutrons: the nucleus converts neutrons into protons and emits a beta particle.

neutron → proton^+ + β^-

Too many protons: the nucleus converts protons into neutrons and emits a positron particle.

proton^+ → neutron + β^+

Too many protons and neutrons: the nucleus emits an alpha particle, which is two neutrons and two protons.

Too much energy: the nucleus releases energy in the form of short-wavelength electromagnetic radiation, gamma rays.

In the first three cases, the nucleus attempts to reach the curve of stable isotopes.

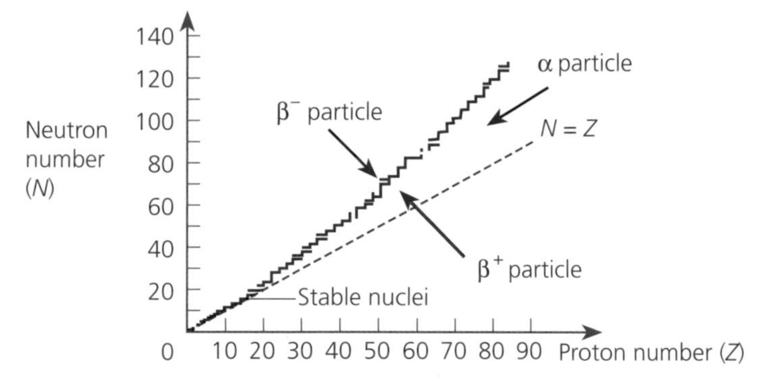

7.1 Properties of radioactive particles

Alpha particles. These are the same as the nucleus of the most common isotope of helium: two protons and two neutrons.

Properties
- Charge $+2e$
- Energy up to 5 MeV
- Strong ionisation
- Range 10 cm in air
- Deflected by magnetic fields

Example reaction

$^{238}_{92}\text{U} \rightarrow {}^{234}_{90}\text{Th} + {}^{4}_{2}\alpha$

Mass number A decreases by four

Atomic number Z decreases by two

Beta particles. These are high-speed electrons.

Properties
- Charge $-1e$
- Energy up to 0.5 MeV
- Medium ionisation
- Range a few millimetres in Al
- Deflected by magnetic fields

Example reaction

$^{60}_{27}\text{Co} \rightarrow {}^{60}_{28}\text{Ni} + {}^{0}_{-1}\beta$

Mass number is unchanged

Atomic number Z increases by one

Positron particles. Same as for beta particles but with a positive charge.

Properties
- Charge $+1e$
- Energy up to 0.5 MeV
- Medium ionisation
- Range a few millimetres in Al
- Deflected by magnetic fields

Example reaction

$^{22}_{11}\text{Na} \rightarrow {}^{22}_{10}\text{Na} + {}^{0}_{+1}\beta$

Mass number is unchanged

Atomic number Z decreases by one

Gamma rays. These are short-wavelength electromagnetic waves.

Properties
- Charge zero
- $E = hf$, energy up to 1 MeV
- Weak ionisation
- Range a few centimetres in lead or metres in concrete
- Not deflected by magnetic fields

Example reaction
$^{60}_{28}\text{Ni} \rightarrow\ ^{60}_{28}\text{Ni} + \gamma$
No change in A or Z

7.2 Detection of radioactive particles

Radioactive particles are usually detected by the ionisation they produce.

Different particles can be distinguished by:
- their deflections in a magnetic field
- their range in different materials

The direction of the force is given by the left-hand rule (see Topic 8, Section 6.1).

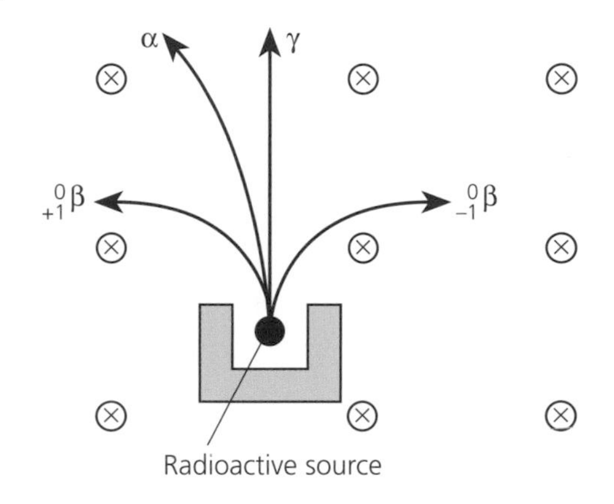

Methods used

Photographic film. Radioactive particles expose photographic film as they pass through it.

Cloud chambers. A condensing liquid forms droplets at points where the air is ionised after the passage of radioactive particles.

Bubble chambers. These are similar to cloud chambers except that when the pressure is reduced in a suitable liquid, such as liquid hydrogen, bubbles form along the ionised path of the particle.

Ionisation chambers and Geiger counters. The ionisation in a gas can be detected and a current pulse generated by the application of a suitable voltage to electrodes. In a Geiger counter, amplification of the ionisation within the gas takes place, producing a larger electrical pulse.

Scintillation counters. Radioactive particles produce light flashes in suitable crystals. The light is detected and amplified by photomultiplier tubes.

7.3 Thickness monitoring

See Section 8.1 for a definition of activity.

Since radioactive particles have mean ranges in materials, this can be used to monitor the thickness of materials by placing a source of known activity on one side of the material and monitoring the activity on the other side. The reduction in activity is related to the thickness of the material.

7.4 Hazards

Radioactive particles originate from two sources:

1. **Natural radioactivity** — cosmic rays, radioactive rocks and our own bodies.
2. **Man-made radioactivity** — X-ray machines, nuclear reactors, fall-out from bombs.

Our bodies can accept a certain level of radiation without any harm.

All radioactive particles entering the body are a hazard, and unnecessary exposure should be avoided. Radioactive particles passing through the body can cause cell death and cell mutations by modifying the DNA, leading to cancer. Some repair does take place.

Precautions

Limit the exposure of the human body to radioactive sources, reducing the dose of radiation received as much as possible, by the following methods:

- shield radioactive sources with suitable thickness of materials, such as lead or concrete, or place the source a suitable distance away
- take care not to ingest radioactive materials
- reduce the exposure time to radioactive sources to a minimum
- since some repair takes place, it is better to be exposed for a long time to a low activity source than for a short time to a high activity source

You should now know:

- **why nuclei are radioactive and the changes in the nucleus when particles are emitted**
- **the nature of the various particles emitted and how they are detected**
- **the hazards associated with using radioactive sources and the methods used to reduce the hazard**

8 *Laws of radioactive decay*

8.1 Activity

A sample that is radioactive emits particles: alpha, beta or gamma. The number of particles emitted per second or the number of atoms decaying per second is a measure of the radioactivity of the sample.

Activity is measured in **becquerels (Bq)**: 1 Bq = 1 decay second^{-1}

μCi = 10^{-6} Ci
mCi = 10^{-3} Ci

A historical unit of activity is the **curie (Ci)**: 1 Ci = 3.7×10^{10} Bq

Radioactivity is a statistical process and the decay of an atomic nucleus is random in that:

- **it is independent of the physical surroundings**
- **we cannot predict which atom will decay**
- **we cannot predict the time at which an atom will decay**

The rate of emission of particles — the activity — also decreases exponentially with time.

The number of undecayed atoms is not constant but decreases exponentially with time.

8.2 Decay constant

The number of atoms dN decaying in a small interval of time dt is proportional to the number of undecayed atoms N in the sample:

$$-\frac{dN}{dt} \propto N$$

$$\frac{dN}{dt} = -\lambda N$$

λ is the **decay constant.**

The same equation can be written in terms of activity:
$A = A_0 e^{-\lambda t}$
A_0 is the activity at time $t = 0$
A is the activity at time t.

Decay equation

Integrating the above equation gives:

$$N = N_0 e^{-\lambda t}$$

where N_0 is the number of undecayed atoms at a time $t = 0$, and N is the number at time t.

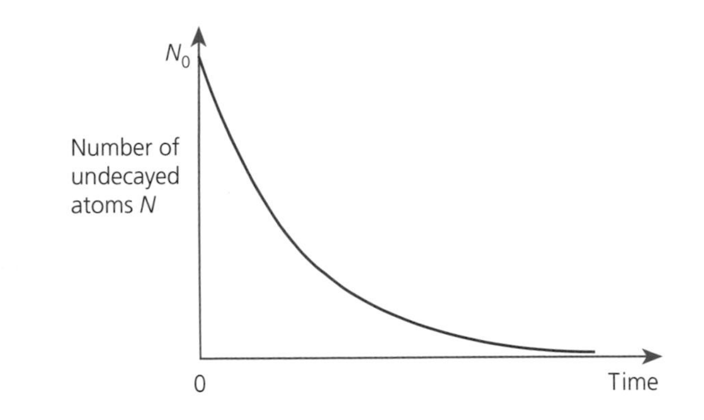

A graph of ln N versus time gives a straight-line graph.

8.3 Half-life

The half-life, $t_{\frac{1}{2}}$, is the time taken for the number of undecayed atoms to fall to half the initial value:

$$\frac{N_0}{2} = N_0 \, e^{-\lambda t_{\frac{1}{2}}}$$

or the activity to fall to half the initial activity.

$$\frac{A_0}{2} = A_0 \, e^{-\lambda t_{\frac{1}{2}}}$$

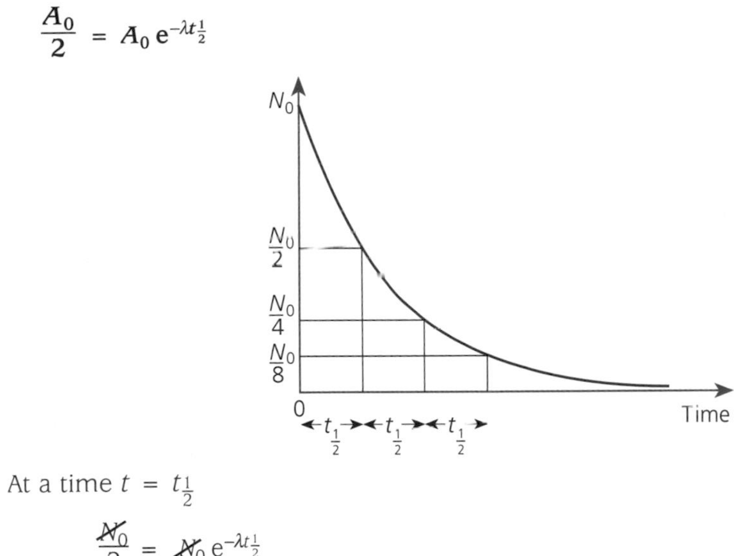

At a time $t = t_{\frac{1}{2}}$

$$\frac{\cancel{N_0}}{2} = \cancel{N_0} \, e^{-\lambda t_{\frac{1}{2}}}$$

$$\frac{1}{2} = e^{-\lambda t_{\frac{1}{2}}}$$

$$2 = e^{\lambda t_{\frac{1}{2}}}$$

$$\ln 2 = \lambda t_{\frac{1}{2}}$$

Try this on your calculator.

Hence half-life and decay constant are related by the equation:

$$\mathbf{0.693 = \lambda t_{\frac{1}{2}}}$$

Note that after a time equal to n half-lives, the number of undecayed nuclei $= (\frac{1}{2})^n N_0$ or the activity $= (\frac{1}{2})^n A_0$.

After one half-life: $N = \frac{N_0}{2}$

After a second half-life: $N = \frac{1}{2}\left(\frac{N_0}{2}\right)$

$$N = \frac{1}{2} \times \frac{1}{2} N_0 = \left(\frac{1}{2}\right)^2 N_0$$

Mean lifetime

We can introduce the idea of a mean lifetime in the same way as for the decay of charge in a capacitor (see Topic 6, Section 6).

The mean life τ of a radioactive sample is the time taken for the number of undecayed atoms or the activity to fall to 37% of the initial value.

$$0.37 N_0 = N_0 e^{-\lambda\tau}$$

$$\ln 0.37 = -\lambda\tau$$

$$-1 = -\lambda\tau$$

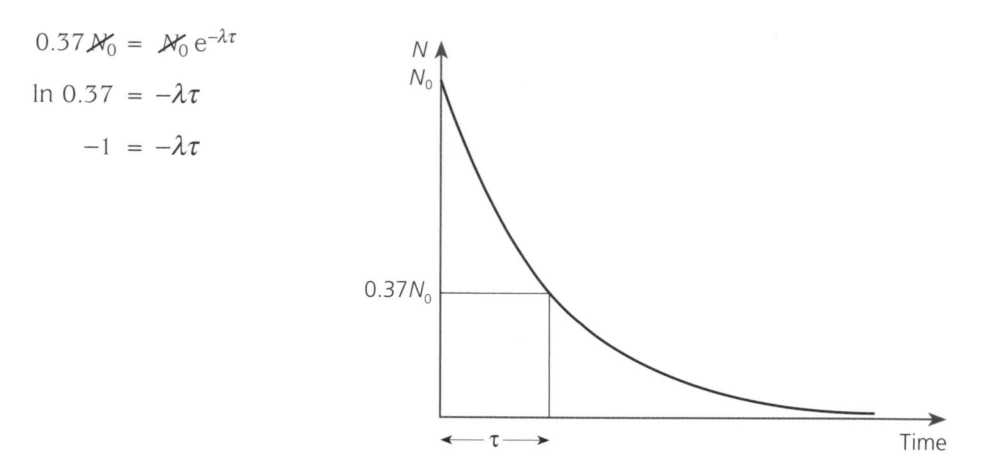

Hence mean lifetime and decay constant are related by the equation

$$\tau = \frac{1}{\lambda}$$

8.4 Radioactive carbon dating

$^{14}_{6}C$ is radioactive with a half-life of 5730 years and combines with oxygen to form CO_2. All living organisms, plants and animals exchange CO_2 while alive, a process that stops when they die. When alive, there is a fixed ratio between $^{14}_{6}C$ and the most common isotope $^{12}_{6}C$. When organisms die, the ratio changes owing to the decay of $^{14}_{6}C$. By measuring the ratio it is possible to estimate the time elapsed since the organism died.

Worked examples

Q1 $^{59}_{28}Fe$ is radioactive and has a half-life of 46 days. Calculate the half-life in seconds, the decay constant and the mean lifetime.

$$t_{\frac{1}{2}} = 46 \times 24 \times 60 \times 60 = 3.97 \times 10^6 \text{ s}$$

$$\lambda = \frac{0.693}{t_{\frac{1}{2}}} = \frac{0.693}{3.97 \times 10^6} = 1.7 \times 10^{-7} \text{ s}^{-1}$$

$$\tau = \frac{1}{\lambda} = 5.7 \times 10^6 \text{ s}$$

Q2 If the half-life of $^{210}_{82}Pb$ is 19 years, how long will it take for the activity to fall to $\frac{1}{16}$ the initial activity?

$$A = \left(\frac{1}{2}\right)^n A_0 \qquad \left(\frac{1}{2}\right)^n = \frac{1}{16} \qquad \therefore n = 4$$

$$\text{time} = 4 \times t_{\frac{1}{2}} = 76 \text{ years}$$

Q3 Strontium-90 has a decay constant of $7.75 \times 10^{-10}\ s^{-1}$. How long would it take for the activity to fall to 70% of the initial value?

$$A = A_0\, e^{-\lambda t}$$

$$70 = 100\, e^{-7.75 \times 10^{-10} t}$$

$$\ln 0.7 = -7.75 \times 10^{-10}\, t$$

$$-0.357 = -7.75 \times 10^{-10}\, t$$

$$t = 4.60 \times 10^{8}\ s$$

You should now know:

- **the properties of a random process**
- **the definition of activity, decay constant, half-life and mean lifetime**
- **the equations of activity versus time, and the number of undecayed atoms versus time in a radioactive sample**

9 Special relativity

Light is a transverse electromagnetic wave. As a result of the work of Maxwell, light was assigned a velocity c with respect to an all-pervading medium called the ether.

9.1 The Michelson–Morley experiment

The Michelson–Morley experiment was designed to measure the velocity of the Earth through the ether.

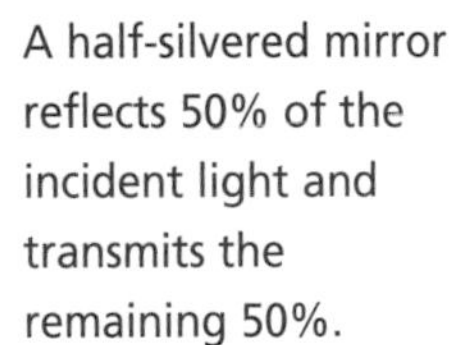
A half-silvered mirror reflects 50% of the incident light and transmits the remaining 50%.

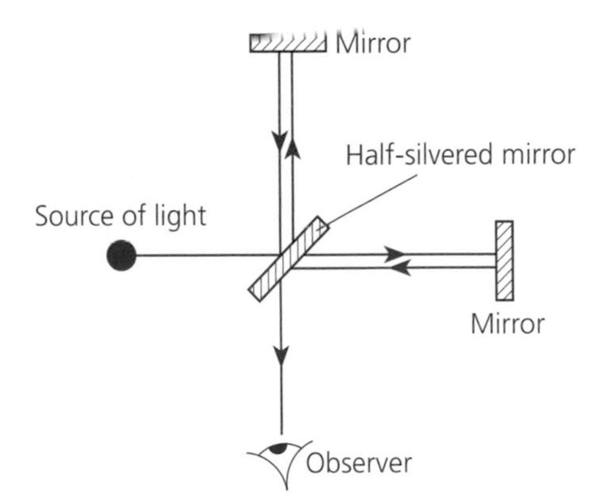

A light beam is split into two beams that travel in perpendicular directions. They are reflected and recombine at the first mirror to form a fringe pattern. The whole system is rotated through 90°. If the system were moving with respect to the ether, there would be a shift in the fringes. No shift is observed.

9.2 Special theory of relativity

These two postulates suggest that there is no absolute frame of reference in which c is measured, hence the null result observed in the Michelson–Morley experiment.

Einstein's postulates

1. **All physical laws have the same form in all inertial frames.**
2. **The speed of light in free space is the same in all inertial frames. It does not depend on the motion of the source or the observer.**

9.3 Time dilation

A pulse of light is sent to a mirror and back and the time of flight is recorded.

Consider an experiment in which a pulse of light is sent to a mirror and reflected. Proper time is the time between two events when an observer is stationary with respect to the two events.

$$\text{total time of flight } T_0 = \frac{2L_0}{c}$$

When the above system is moving with a velocity with respect to an observer, the velocity of light is still c but time in the moving frame is slower. T is the time of flight observed.

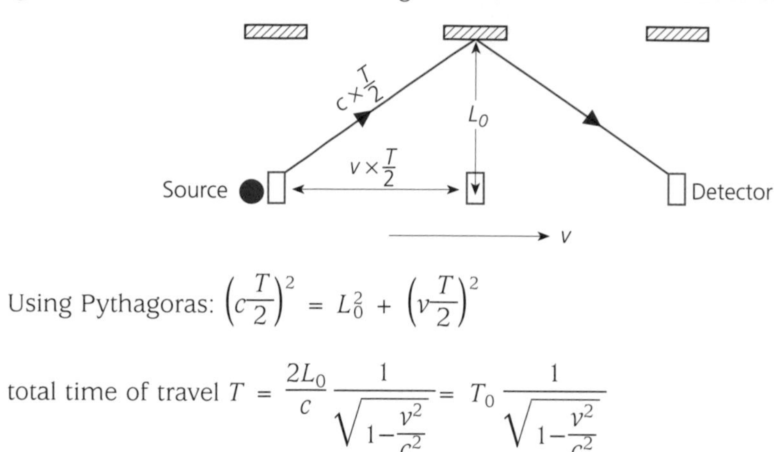

$$\text{Using Pythagoras: } \left(c\frac{T}{2}\right)^2 = L_0^2 + \left(v\frac{T}{2}\right)^2$$

$$\text{total time of travel } T = \frac{2L_0}{c}\frac{1}{\sqrt{1-\frac{v^2}{c^2}}} = T_0\frac{1}{\sqrt{1-\frac{v^2}{c^2}}}$$

T is greater than T_0 and this is called **time dilation**.

Muon decay

Muon velocity = 0.995c.
The mean lifetime observed when stationary with respect to an observer = 2.2 μs.

Muons arrive in cosmic rays from outer space and have velocities close to that of light. The experiment measured the initial number of muons N_0 at a height of 2000 m and the number N at sea level. The number detected at sea level was much greater than their stationary mean lifetimes predicted. Their mean lifetimes had been extended due to time dilation.

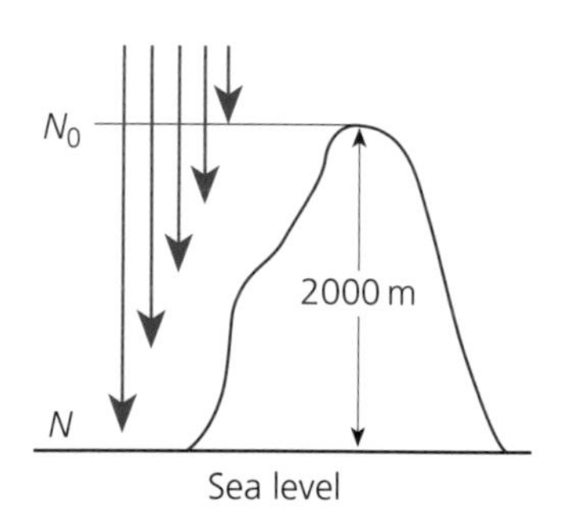

9.4 Length contraction

Now consider making measurements while travelling alongside the muons. The mountain is now in a moving frame of reference. The muons see the height of the mountain contracted from 2000 m so that the number arriving at the base is the same as in the experiment above. This is called **length contraction**.

A length L_0 in a frame moving with velocity v is measured as L where:

$$L = L_0\sqrt{1-\frac{v^2}{c^2}}$$

9.5 Mass energy

A very small mass is equivalent to a large amount of energy. 1 kg of apples is equivalent to $1 \times (3 \times 10^8)^2 = 9 \times 10^{16}$ J, the energy contained in 1.5×10^9 litres of petrol.

One other consequence of the special theory of relativity is that mass and energy are equivalent and related by the equation:

$$E = mc^2$$

where m is the mass in kilograms and c is the velocity of light.

This equation is used to explain how energy is obtained in nuclear fission and fusion (see Section 10), where the final mass of the products in both reactions is less than the initial mass. The mass lost is converted into energy, as required by the above equation.

9.6 Relativistic mass

As an object gains energy, its mass is found to increase according to the equation:

$$m = \frac{m_o}{\sqrt{1-\frac{v^2}{c^2}}}$$

where m_0 is the mass of the object at rest.

The energy of an object with rest mass m_0 is:

$$E = mc^2 = \frac{m_o c_2}{\sqrt{1-\frac{v^2}{c^2}}}$$

At velocities close to c the mass increase is significant. An electron accelerated to a velocity 0.99c has a mass seven times its mass at rest.

Kinetic energy is energy associated with an increase in mass of the object. At velocities v much less than c, the above equation reduces to the familiar kinetic energy equation:

$$E = \frac{1}{2}mv^2$$

9.7 Binding energy of the nucleus

When individual particles come together to form a nucleus, energy is released which is called the binding energy. This energy results from the fact that the mass of the nucleus is less than the mass of the individual protons and neutrons from which the nucleus is made.

See Section 6.1 for the definition of u.

Consider the most abundant isotope of iron, $^{56}_{26}Fe$, which has a mass number of 56 made up of 26 protons and 30 neutrons, and 26 electrons. The measured mass of this atom is 55.93493 u.

The total mass of the individual particles

	Number	Mass	Total mass
Neutrons	30	1.00867	30.26010
Protons	26	1.00728	26.18928
Electrons	26	0.00055	0.01430
		Total mass	56.46468

Compare this value with the measured mass of an atom of $^{56}_{26}Fe$, 55.93493u, a difference of 0.52875 u. This is called the **mass defect**.

All atoms and their isotopes have a mass defect and, when the individual particles come together to form a nucleus, energy equivalent to the mass defect is released. This is called the **binding energy of the nucleus**.

In the case of $^{56}_{26}Fe$, the energy released $= 0.529 \times 1.66 \times 10^{-27} \times (3.00 \times 10^8)^2$

$$= \frac{7.90 \times 10^{-11}}{1.60 \times 10^{-19}} \text{ eV}$$

So the binding energy $= 4.94 \times 10^8$ eV $= 494$ MeV

Remember, 1 u $= 1.66 \times 10^{-27}$ kg.

See Section 10.1 for the application of this.

9.8 Binding energy per nucleon

A quantity often used in nuclear physics is the binding energy per nucleon, the total binding energy divided by the number of nucleons (protons plus neutrons) in the nucleus.

So the binding energy per nucleon of $^{56}_{26}\text{Fe} = \frac{494}{56} = 8.8$ MeV per nucleon.

Note that the energy released is equivalent to the mass defect.

Worked examples

Q1 Muons have a mean lifetime of 2.2 μs and a velocity of 0.995*c*. What fraction of those found at a height of 2000 m would be expected to reach the ground assuming no time dilation takes place?

$$N = N_0 e^{-\lambda t} = N_0 e^{-\frac{t}{\tau}}$$

$$\text{time of flight} = \frac{2000}{0.995\,(3 \times 10^8)} = 6.7 \times 10^{-6}\ \text{s}$$

$$N = N_0 e^{-\frac{6.7 \times 10^{-6}}{2.2 \times 10^{-6}}}$$

$$\frac{N}{N_0} = 0.047$$

What would be the answer assuming time dilation occurs?

$$\text{New } \tau \text{ due to time dilation: } \tau = \frac{2.2 \times 10^{-6}}{\sqrt{1 - 0.995^2}} = 2.2 \times 10^{-5}$$

$$N = N_0 e^{-\frac{6.7 \times 10^{-6}}{2.2 \times 10^{-5}}} = N_0 e^{-0.3}$$

$$\frac{N}{N_0} = 0.74$$

Q2 How much energy is released when a proton and an antiproton collide?

total mass of a proton and an antiproton = 2 × 1.0073 u

energy = $mc^2 = 2 \times 1.0073 \times 1.6606 \times 10^{-27} \times (3.00 \times 10^8)^2 = 3.01 \times 10^{-10}$ J

Q3 $^{238}_{92}\text{U}$ has a rest mass of 238.05080 u. Calculate the mass defect and the binding energy of $^{238}_{92}\text{U}$.

$92 \times 1.00728 = 92.66976$

$92 \times 0.00055 = 0.05060$

$146 \times 1.00867 = 147.26582$

$= 239.98618$ u

239.98618 − 238.05080

mass defect = 1.93538 u

binding energy = $1.93538 \times 1.6606 \times 10^{-27} \times (3.00 \times 10^8)^2$

$= 2.89 \times 10^{-10}$ J

$$= \frac{2.89 \times 10^{-10}}{1.60 \times 10^{-19}} = 1.81 \times 10^9\ \text{eV} = 1810\ \text{MeV}$$

You should now know:

- **Einstein's postulates of special relativity**
- **the concept of time dilation and length contraction**
- **the muon experiment**
- **the equivalence of mass and energy**
- **the definition of mass defect and binding energy per nucleon**

10 Fission and fusion

10.1 Nuclear binding energy

The particles in the nucleus — protons and neutrons — are held together by strong attractive nuclear forces. In order to break a nucleus apart into individual particles, energy must be supplied to do work against the attractive forces. This energy is called the **nuclear binding energy**.

Another way of looking at the nuclear binding energy is to say that it is the **energy given out** when isolated protons and neutrons come together to form a nucleus.

The mass of a nucleus is less than the total mass of the individual protons and neutrons. The binding energy released is this mass loss, Δm, converted into energy using energy = Δmc^2.

$$\textbf{binding energy per nucleon} = \frac{\textbf{the total binding energy for a given nucleus}}{\textbf{the number of nucleons in the nucleus}}$$

Binding energy per nucleon is not constant but has the curve shown below.

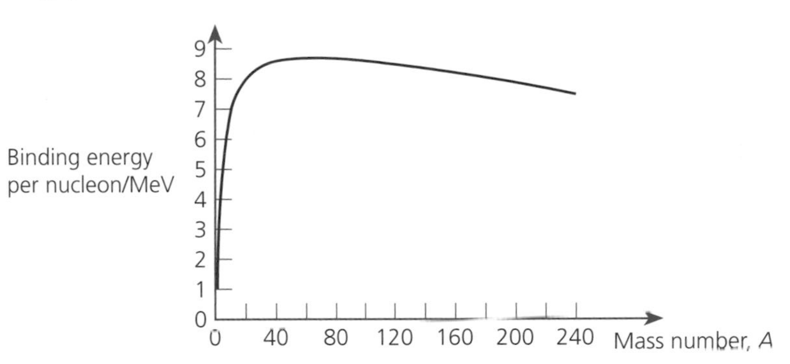

It is the shape of this curve that gives rise to the emission of energy in nuclear fission and fusion.

10.2 Nuclear fission

This is the process in which a $^{238}_{92}U$ nucleus splits into two lighter nuclei. In a nuclear reactor, the fission process is continuous as follows:

1. When a thermal neutron collides with a $^{238}_{92}U$ nucleus, it splits into two lighter nuclei, X and Y. High-speed neutrons and energy are released.
2. These neutrons are then slowed down to thermal energies by a moderator, typically graphite or water, in the reactor.
3. When slowed to thermal velocities, if at least one neutron collides with another $^{238}_{92}U$ nucleus causing it to split into two with the emission of some fast neutrons, the process continues indefinitely.

If on average more than one neutron causes a new fission, the process escalates; if less than one neutron causes a new fission, the process slows down and eventually stops.

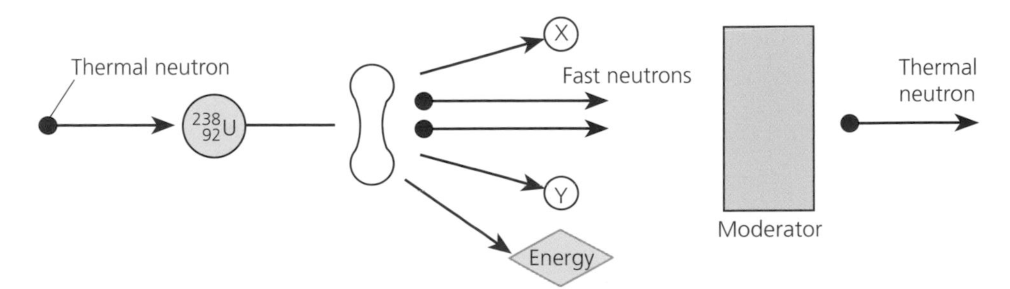

This is called a **chain reaction**.

Nuclear fission is controlled by inserting rods of boron into the reactor. Boron absorbs neutrons so that they cannot cause further fissions.

This equal split is an ideal case and, in practice, a range of nuclear masses and hence atoms are produced in nuclear fission.

10.3 Energy released during fission

Consider the following simple fission process in which one $^{238}_{92}U$ nucleus splits into two nuclei of equal mass, $^{118}_{46}Pd$, and emits two fast neutrons.

The binding energy per nucleon for $^{238}_{92}U$ is about 7.6 MeV per nucleon; for $^{118}_{46}Pd$ it is about 8.3 MeV per nucleon (approximate numbers).

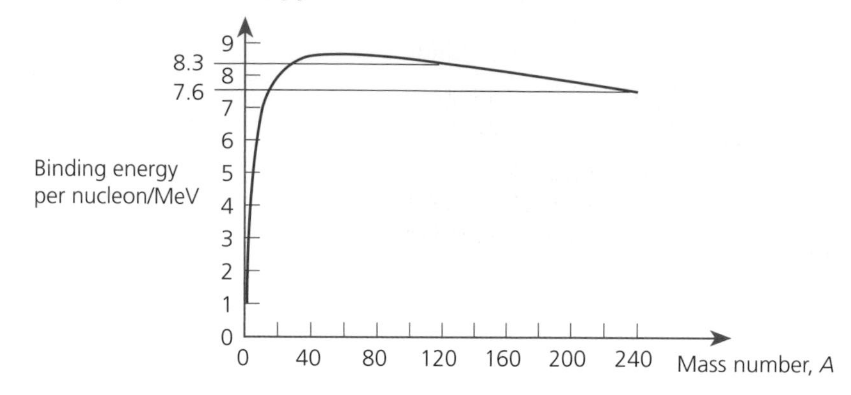

The energy released when a $^{238}_{92}U$ nucleus is created from protons and neutrons is:

$$238 \times 7.6 = 1809 \text{ MeV}$$

In this simple calculation, we have not considered the two neutrons also released in the reaction.

The energy released when two $^{118}_{46}Pd$ nuclei are formed as a result of fission is:

$$2 \times 118 \times 8.3 = 1959 \text{ MeV}$$

The extra energy released by the fission process is therefore:

$$1959 - 1809 = 150 \text{ MeV}$$

Hence, in moving from one $^{238}_{92}U$ nucleus to two $^{118}_{46}Pd$ nuclei, extra energy of approximately 150 MeV is released.

Energy is released as a result of the fission because the product nuclei have a higher binding energy per nucleon.

10.4 Nuclear fusion

Nuclear fusion is the source of energy in the Sun.

If we join together nuclei at the bottom of the periodic table to produce a heavier atom, again the binding energy per nucleon of the product nucleus is greater than the parent, and energy is released in the process. This is **nuclear fusion**.

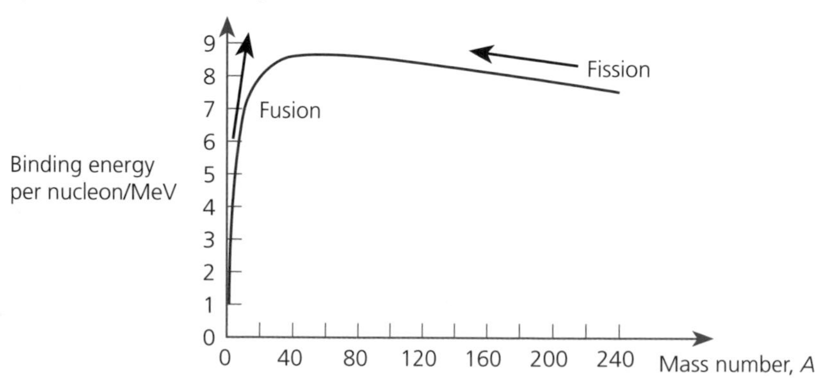

To make two nuclei join together, we must make them collide with very high energy, so that they are close enough together for the short-range nuclear attractive force to take effect. Remember: as they move together, the coulombic repulsion grows dramatically.

^2_1H is an isotope of hydrogen. 0.01% of sea water contains ^2_1H, so the fuel for fusion is widely available.

A simplified fusion process

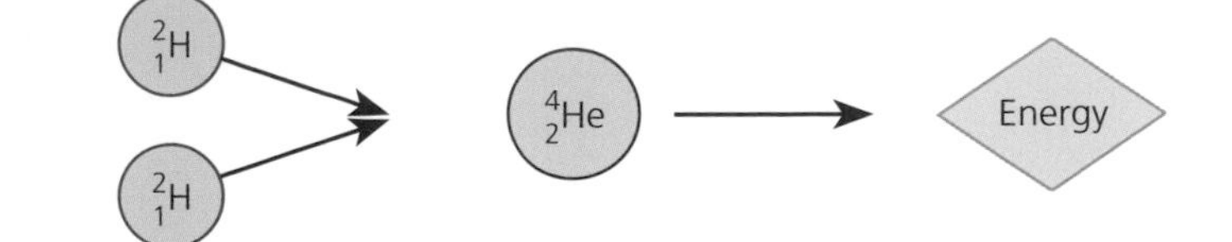

A practical fusion reaction involves several stages intermediate to the above.

See Topic 4, Section 4, the kinetic theory of gases.

Two conditions, which are hard to achieve, are required for fusion to take place:

1. **The particles in the gas must have very high energies to overcome the coulombic repulsion. Hence, very high temperatures are required, (50 000 000 K).**
2. **The particles must have a very high density to increase the probability of suitable collisions between nuclei.**

Fusion machines are currently being developed to generate controlled fusion reactions.

Under these conditions, the gas particles are ionised, and the gas is called a **plasma**. These very hot gas plasmas are confined to doughnut rings using magnetic and electric fields in machines called Tokamaks.

Comparison of fission and fusion

The atoms produced by nuclear fission contain too many neutrons for stability (see Section 7). They are therefore radioactive, emitting beta particles and gamma rays — some with half-lives of thousands of years. So the waste products of nuclear fission are dangerous and must be stored safely.

In comparison, the atoms resulting from nuclear fusion are helium, which is an inert gas and non-radioactive.

Worked examples

Q1 ^4_2He has an atomic mass of 4.0026 u. Calculate the binding energy per nucleon in joules and MeV.

$$2 \times 1.00728 = 2.01456$$
$$2 \times 0.00055 = 0.00110$$
$$2 \times 1.00867 = 2.01734$$
$$\overline{4.03300 \text{ u}}$$

$$4.03300 - 4.00260 = 0.03040 \text{ u} = \text{mass defect}$$

$$\text{binding energy} = 0.0304 \times 1.66 \times 10^{-27} \times (3.00 \times 10^8)^2$$
$$= 4.54 \times 10^{-12} \text{ J}$$
$$= \frac{4.54 \times 10^{-12}}{1.60 \times 10^{-19}} = 2.84 \times 10^7 \text{ eV} = 28.4 \text{ MeV}$$

$$\text{binding energy per nucleon} = \frac{28.4}{4} = 7.1 \text{ MeV per nucleon}$$

Q2 In the following radioactive decay:

$$^{238}_{92}U \rightarrow {}^{234}_{90}Th + {}^4_2He + \text{energy}$$

alpha particles are released with a kinetic energy that comes from the reduction in total mass of the nuclei as a result of the reaction. The binding energy per nucleon of $^{238}_{92}U$ = 7.567 MeV, of $^{234}_{90}Th$ = 7.594 MeV and of ^4_2He = 7.075 MeV. Calculate the energy with which the alpha particle is released.

total binding energy of $^{238}_{92}U$ = 7.567 × 238 = 1801 MeV

total binding energy of $^{234}_{90}Th$ = 7.594 × 234 = 1777 MeV

total binding energy of $^{4}_{2}He$ = 7.075 × 4 = 28.30 MeV

total of products = 1777 + 28.30 = 1805 MeV

energy of the α particle = 1805 – 1801 = 4 MeV

Q3 In a theoretical fusion reaction, two atoms of deuterium, $^{2}_{1}H$, are fused to form one atom of helium, $^{4}_{2}He$. If the binding energy per nucleon of deuterium is 1.1 MeV and the binding energy of helium is 7.1 MeV, how much energy, in joules, would be released in each fusion?

total binding energy of $^{4}_{2}He$ = 4 × 7.1 = 28.4 MeV

total binding energy of 2 × $^{2}_{1}H$ = 2 × 2 × 1.1 = 4.4 MeV

energy released = 24.0 MeV

energy in joules = $24.0 \times 1 \times 10^{6} \times 1.6 \times 10^{-19} = 3.8 \times 10^{-12}$ J

You should now know:
- **the definition of binding energy and binding energy per nucleon**
- **the fission chain reaction**
- **the magnitude of the energy released per fission**
- **nuclear fission and fusion and the binding energy per nucleon curve**
- **the simple fusion process**
- **the advantages of nuclear fusion over nuclear fission**

11 *Elementary particles*

Until about 1932, it was thought that there were four elementary particles: the electron, the proton, the neutron and the photon. In 1932, a new particle, the positron (see Section 7), was discovered, followed soon after by the muon and then the pion, and an array of new particles. The majority of these particles result from high-energy collisions at GeV energies in particle-accelerating machines. Many of the particles observed exist only for very short lifetimes. In attempting to explain the vast array of particles, we use the standard model in which the particles can be considered to be constructed from two groups of particles: **leptons** and **quarks**.

1 GeV = 1×10^{9} eV

11.1 Leptons, baryons and mesons

Known particles fall into one of three groups: leptons, baryons and mesons. They are distinguished by:
- their charge as a multiple of the charge on an electron
- their mass, usually quoted as the energy equivalent using the conversion $E = mc^2$
- their spin, as many particles possess spin rather like a top, in units of $\frac{h}{2\pi}$

Leptons. There are six leptons, including the electron, e, and the muon, μ, which has the same charge as an electron but 200 times the mass. Neutrinos are also in this group. They have zero mass and zero charge and are recognised by their spin.

Large numbers of muons are created at the top of the Earth's atmosphere by cosmic rays.

Baryons. The most numerous group, containing such particles as protons, neutrons plus numerous others, given Greek capital letters Δ, Σ, Ω etc. Baryons have spins, in units of $\frac{1}{2}\frac{h}{2\pi}$ and $\frac{3}{2}\frac{h}{2\pi}$.

Mesons. This is another numerous group with particles such as pions, π, σ, ψ etc. The masses of mesons and baryons are similar, in the range 140 MeV to 3700 MeV, but mesons have spins in units of $\frac{h}{2\pi}$ and $\frac{2h}{2\pi}$.

Antiparticles. All the particles listed above have antiparticles, which have an opposite charge but the same mass and spin.

Antiparticles are distinguished by a bar over the top. For example, e is an electron and $\bar{e}$ is the antielectron.

When a particle meets its antiparticle, they annihilate each other and release energy.

Interacting forces

All the forces in nature can be explained in terms of four basic interactions:

- strong nuclear interaction
- electromagnetic interaction
- weak nuclear interaction
- gravitational interaction

Note that there are two nuclear interactions: a weak and a strong interaction.

The **weak interaction** takes place between electrons or positrons and nucleons in the nucleus. The weak interaction is also observed in beta and positron decay.

The **strong interaction** is the force that holds the nucleus together and exists between baryons and mesons. Particles that interact by the strong force are called hadrons, so baryons and mesons are both hadrons.

11.2 Interactions between particles — conservation laws

When high-energy particles interact in particle-accelerating machines, other particles are created. Some of these particles decay to produce others. The particles created in any interaction or decay are governed by conservation laws.

Some of these conservation laws will be familiar from particle collisions (Topic 4, Section 3):

- **the total momentum must be conserved**
- **the total mass/energy must be conserved**

Additional conservation rules are:

- **the total charge must be the same before and after the interaction**
- **the net baryon number must be the same before and after the interaction**
- **the net lepton number must be the same before and after the interaction**

Baryons have a baryon number 1 and antibaryons −1. Leptons have a lepton number 1 and antileptons −1.

There are certain conservation rules that apply only to strong nuclear interactions — those between hadrons. One of these rules is **strangeness**. Each hadron is given a **strangeness number** and strangeness must be conserved in an interaction.

11.3 Quark model

Leptons appear to be fundamental particles in that they do not break down into smaller particles. Hadrons are more complex and decay into other hadrons. The quark model states that hadrons are composed of two or three elementary particles: quarks. There are now six quarks — up, down, strange, charmed, top and bottom. Each quark has an antiquark. Quarks have charges — $-\frac{1}{2}$ or $+\frac{2}{3}$ the charge on an electron — and a baryon number — $+\frac{1}{3}$ for quarks and $-\frac{1}{3}$ for antiquarks.

A proton is two up quarks and one down quark; a neutron is two down quarks and one up quark.

Baryons are constructed from three quarks; mesons are constructed from a quark and an antiquark.

11.4 Fields

We have already looked at fields in terms of a force acting at a distance. An electron in the electric field experiences a force due to another electron, and the magnitude of the field is measured in terms of the force on a unit charge. We must now look at what gives rise to this force.

In quantum theory, energy is found in discrete units or quanta. Hence, the energy in fields must also be in small packets of energy quanta.

Field	Quanta	Mass/MeV
Gravitational	Gravitons	0
Weak	W and Z particles	82 000 and 93 000
Electromagnetic	Photons	0
Strong	Pions and mesons	140

A field associated with a particle can be pictured as quanta being emitted and absorbed by the particle and, when there is interaction with a second particle, the interaction can be explained in terms of an exchange of the relevant quanta.

Two electrons experience a force of repulsion by the exchange of photons.

These interactions can be drawn on **Feynman** diagrams, where the position of the electron is plotted as a function of time. The first electron creates a photon, and then the second electron annihilates the photon.

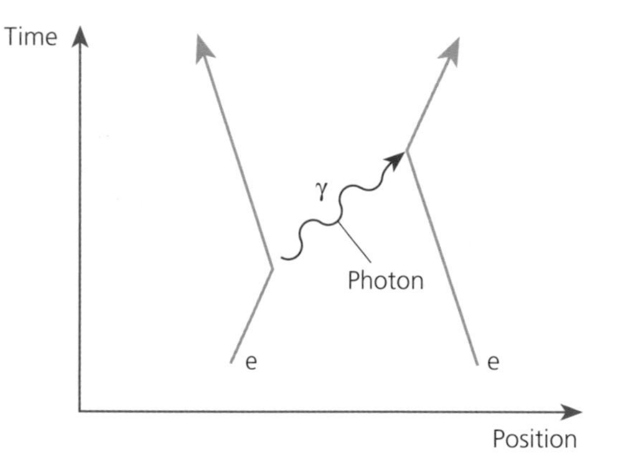

You should now know:

- **the grouping of fundamental particles — leptons, baryons, mesons, hadrons**
- **some of the conservation laws applied to the interaction of fundamental particles**
- **the quark model**
- **fields of force as the exchange of a suitable quanta**

TOPIC 10 Astrophysics

1 Astronomical telescopes

The two main types of optical telescopes used in astronomy are refracting telescopes and reflecting telescopes.

1.1 Refracting telescope

The first lens produces a real image at its focus. The second lens acts a magnifying glass.

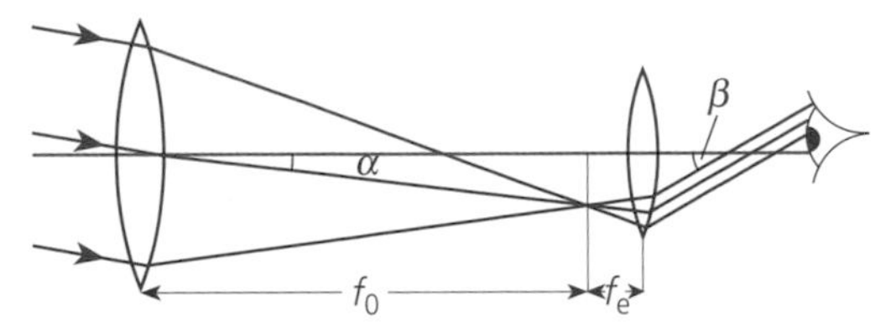

1.2 Reflecting telescope

The first lens is replaced by a concave mirror, which gathers the light.

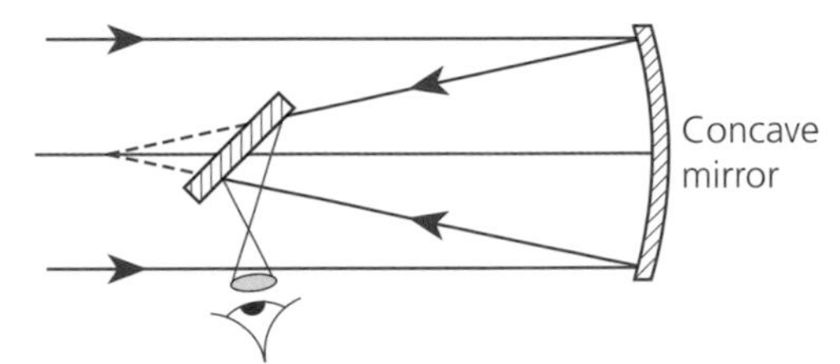

Two factors are important in the design of telescopes:

- the size of the objective lens, or the mirror, as it controls the brightness of the image
- the angular magnification

$$\text{angular magnification} = \frac{\text{angle subtended by the image}}{\text{angle subtended by the image viewed through the eyepiece}}$$

In the case of the reflecting telescope, the angular magnification is calculated as follows:

for small angles, $\tan\beta = \frac{AB}{f_e} \approx \beta$

and $\tan\alpha = \frac{AB}{f_0} \approx \alpha$

angular magnification $= \frac{\alpha}{\beta} = \frac{f_0}{f_e}$

The same equation applies to refracting telescopes.

Thus a large magnifying power requires an objective lens or mirror with a large focal length and an eyepiece with a short focal length.

Refracting telescopes suffer from two major disadvantages:

- the large objective lens required is difficult to manufacture and heavy
- they suffer from chromatic aberration, where different colours do not focus at the same point and a blurred image is produced

In a reflecting system, all colours are reflected through the same angle by the objective mirror.

1.3 Radio telescopes

The design of radio telescopes is similar to that of reflecting telescopes except that the detector of the electrical signals only images a point and the telescope must scan to obtain a two-dimensional radio image.

1.4 Resolving power

Diffraction occurs in telescopes and the ability of a telescope to resolve two objects depends on the overlap of the circular diffraction patterns (see Topic 7, Section 9.3).

θ is the angular separation of the two images that are just resolved.

$$\theta = \frac{1.22\lambda}{D}$$

where D is the diameter of the objective lens or mirror.

Radio telescopes use the same equation but since the wavelength of radio waves is very large, even with large dishes the resolution is much less than with optical telescopes.

You should now know:

- **the construction of refracting and reflecting telescopes**
- **how to calculate the angular magnification**
- **the criterion for the resolution of telescopes**

2 *Astronomy*

2.1 Structure of the universe

The universe is composed of:

- the gases hydrogen and helium
- stars as represented by our own Sun and planets with at least 20% of stars as double (binary) stars
- spiral galaxies, global clusters and elliptical galaxies
- elements in the form of gas/dust up to atomic number 92
- radiation
- dark matter

2.2 Measurements on stars

Apparent brightness: the energy entering the eye or telescope per square metre. The energy per unit area decreases with the square of the distance from the star.

Take care — the magnitude scale works backwards. The greater the magnitude, the dimmer the star.

Apparent magnitude (*m*): the classification of a star's brightness into equal brightness steps as perceived by the eye. First-magnitude stars are the brightest and sixth magnitude stars are the dimmest that can be seen. A difference in brightness of five magnitudes corresponds to a ratio of brightness of 100:1.

A difference of one magnitude = $\sqrt[5]{100}$ = 2.512 in brightness.

A second magnitude star is 2.512 times brighter than a third magnitude star and 2.512 × 2.512 times brighter than a fourth magnitude star.

Astronomical unit (AU): the mean distance from the Earth to the Sun.

$1\text{AU} = 1.5 \times 10^{11}$ m

The distance travelled by light in 1 year = $365 \times 24 \times 3600 \times 3 \times 10^8 = 9.5 \times 10^{15}$ m

Light year (cy): the distance travelled by light in 1 year.

2.3 Measurement of stellar distances

It is possible to measure the distances of stars close to our galaxy by the relative movement of the star with respect to the cosmic background as the Earth rotates around the Sun.

c is very small, hence:

$$\tan\frac{c}{2} = \frac{c}{2} = \frac{1\text{AU}}{d}$$

where d is the distance to the star.

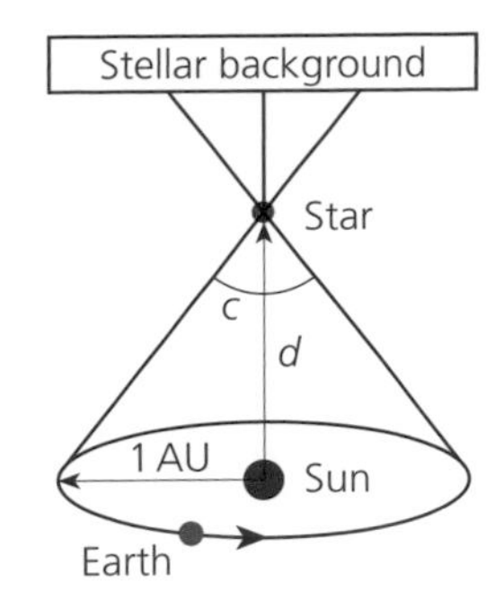

Parsec (pc): in the diagram above, when $\frac{c}{2} =$ 1 sec of an arc, the distance to the star is called 1 parsec.

For a star:
$L = 4\pi d^2 b$.
For the Sun:
$L_\odot = 4\pi d_\odot^2 b_\odot$
where d is the distance to the star/Sun and b is apparent brightness.

Luminosity (L): the total energy radiated in watts by the star, usually expressed in terms of $L_\odot$, the luminosity of the Sun.

$$\frac{L}{L_\odot} = \frac{4\pi d^2 b}{4\pi d_\odot^2 b_\odot} = \frac{d^2 b}{d_\odot^2 b_\odot}$$

Absolute magnitude (M): the apparent magnitude of a star viewed at a distance of 10 pc.

$$m - M = 5\log\frac{d}{10}$$

Worked examples

Q1 Calculate the distance of 1 pc in metres.

$$1'' = \frac{2 \times \pi}{360 \times 60 \times 60}$$

$$1'' = 4.8 \times 10^{-6}\ \text{rad}$$

$$1\text{pc} = \frac{1\text{AU}}{1''}$$

$$1\text{pc} = \frac{1.5 \times 10^{11}}{4.8 \times 10^{-6}}$$

$$1\text{pc} = 3.1 \times 10^{16}\ \text{m}$$

Q2 A distant star is 3.56 pc from the Earth. When viewed from the Earth it appears 5.6×10^{-12} as bright as the Sun. What is the ratio of its luminosity to that of the Sun?

$$3.6\text{pc} = \frac{3.6 \times 3.1 \times 10^{16}}{1.5 \times 10^{11}} = 7.4 \times 10^{5}\ \text{AU}$$

$$\frac{L}{L_\odot} = \frac{d^2 b}{d_\odot^2 b_\odot} = (7.4 \times 10^5)^2 \times 5.6 \times 10^{-12} = 3.1$$

Q3 A star at a distance of 3.6 pc has an apparent magnitude of +4.7. Calculate its absolute magnitude.

$$4.7 - M = 5\log\frac{3.6}{10} = 5 \times (-0.44)$$

$$4.7 - M = -2.2$$

$$M = +6.9$$

You should now know:

- **the structure of the universe**
- **the magnitude scale of brightness**
- **the definition of AU and light year**
- **the measurement of stellar distances**

3 *Classification of stars by temperature*

Stars approximate to black bodies (full radiators) in which the spectrum emitted follows a characteristic shape that depends on the temperature of the body (see Topic 4, Section 8.3).

Wien's displacement law

λ_{max} increases as the temperature decreases.

$$\lambda_{max} = \frac{3 \times 10^{-3}}{T}$$

Stefan's law

energy per second per unit area $= \sigma T^4$

where σ is Stefan's constant.

Hence:

the luminosity of a star = surface area of the star $\times \sigma T^4$

The total energy radiated by a black body is the area under the black body curve.

3.1 Calculation of stellar diameters from radiation laws

1 Measure the luminosity from the apparent brightness and the distance.
2 Obtain the temperature of the star from the black body spectrum and apply Stefan's law.

$$L = 4\pi R^2 \sigma T^4$$

$$R = \sqrt{\frac{L}{4\pi\sigma T^4}}$$

3.2 Stellar spectral classes

The HR diagram

Stars are grouped by plotting their luminosity against their temperature.

The scale of the vertical axis is based on the luminosity of our Sun. On this scale, a star as luminous as our Sun has a value of 1.

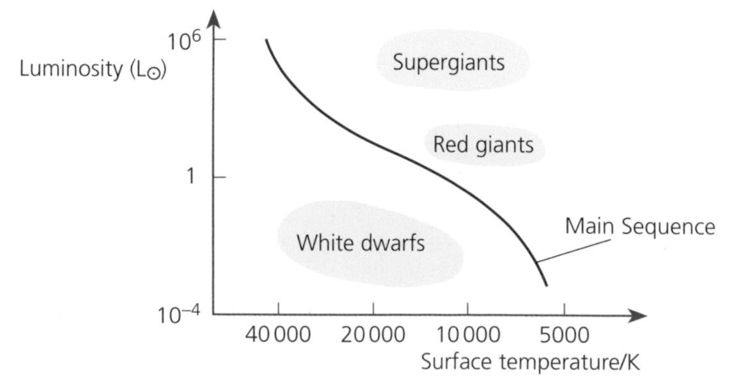

The main sequence

The structure of a star is uniquely determined by its mass and its composition. The most massive stars are the most luminous and the hottest. The diagonal line on the HR diagram represents the main sequence.

Red giants and supergiants

Red giants and supergiants are characterised by:

- high luminosity and low surface temperature
- large surface area, hence large radii

White dwarfs

White dwarfs are characterised by:

- low luminosity and high surface temperature
- small surface area, hence small radii
- very high density

Spectral classes

To remember this use the mnemonic 'Oh be a fine guy/girl, kiss me'.

The black body spectrum of stars contains black absorption lines associated with known terrestrial elements. The stars on the main sequence can thus be grouped into seven spectral classes, identified by the letters O, B, A, F, G, K and M.

The absorption lines in the various classes are mainly due to:

- O, helium
- B and A, hydrogen
- F and G, metals
- K and M, molecules

3.3 Stellar evolution

The first generation of stars began with a composition of pure hydrogen and helium. Other elements were synthesised in their hot centres. Stars such as the Sun with heavy elements in the outer layers must have been formed as second-generation stars.

M_0 is the mass of the Sun, called the solar mass $M_0 = 1.99 \times 10^{30}$ kg.

The ultimate fate of a star depends on its mass. Stars in the main sequence follow essentially the same fusion process, converting hydrogen into helium, and have masses from $0.1M_0$ to $60M_0$.

Stars with mass $< 0.1M_0$ do not have enough material for fusion. Stars with mass $> 60M_0$ are unstable.

For low-mass stars ($< 12M_0$), the cycle is as follows:

- hydrogen is consumed and so helium starts to fuse
- producing larger atoms in the core, which heats up
- the radius increases and the luminosity decreases
- the star becomes a red giant and then a super red giant
- the red giant gives off its outer layers, passing through a planetary nebula stage
- eventually cooling to a white dwarf consisting of helium or higher mass elements

High-mass stars ($> 12M_0$):

- evolve much more quickly
- gravitational pressure causes fusion between heavier elements
- fusion stops when the core material has been fused to iron
- the star collapses in on itself, the core heats up
- this causes a catastrophic explosion called a supernovae
- material is blown away and a central core is left

Neutron stars and pulsars

The diameter of a neutron star is about 12 km; its density is 1.2×10^{17} kg m^{-3}.

When the central core is less than about $3M_0$, a core of neutrons is left, called a neutron star. Regularly pulsing radio sources (pulsars) are thought to be associated with neutron stars as they cool and release energy.

See Topic 8, Section 3.6

Black holes

The velocity of escape for a mass m from a body of mass M and radius R is:

$$v_e = \sqrt{\frac{2GM}{R}}$$

For a neutron star of mass M, the velocity of escape equals the velocity of light when the radius is given by the equation:

$$R_s = \frac{2GM}{c^2}$$

where R_S is the Schwarzschild radius. When the resulting neutron star has a mass $> 3M_0$, this radius is less than R_S, so nothing can escape, including light, and a black hole is created.

Worked examples

Q1 A bright orange star in the constellation Orion has a λ_{max} of 7.1×10^{-7} m and luminosity of 1×10^{20} watts. What is its radius?

$$7.1 \times 10^{-7} = \frac{3.0 \times 10^{-3}}{T}$$

$T = 4200$ K

$$R = \sqrt{\frac{L}{4\pi\sigma T^4}} = \sqrt{\frac{1 \times 10^{20}}{4 \times \pi \times 5.67 \times 10^{-8} \times (4200)^2}} = 2.8 \times 10^9 \text{ m}$$

Q2 Calculate the Schwarzschild radius, in metres, of a black hole with 12 times the mass of the Sun.

mass = $12 \times 1.99 \times 10^{30} = 2.4 \times 10^{31}$ kg

$$R_s = \frac{2GM}{c^2} = \frac{2 \times 6.67 \times 10^{-11} \times 2.4 \times 10^{31}}{(3 \times 10^8)^2} = 35\,600 \text{ m}$$

You should now know:

- **Wien's and Stefan's laws**
- **how to calculate stellar diameters**
- **the HR diagram**
- **red giants and white dwarfs**
- **stellar evolution and the fate of low-mass and high-mass stars**

4 *Cosmology*

Cosmology is the science concerned with the structure and evolution of the universe.

4.1 Oblers's paradox

If the universe were infinite and static, the sky at night would be light because a star would be seen along any line of sight. The sky is dark, so the universe is finite and we can only see light from a galaxy if it is near enough so that light emitted has had time to reach us during the life of the universe.

Doppler effect

$$\frac{\lambda - \lambda'}{\lambda} = \frac{\lambda' - \lambda}{\lambda} = \frac{v}{c}$$

$\lambda - \lambda'$ is the shift in wavelength and v is the relative velocity between the source and the observer.

4.2 Hubble's law

Hubble observed that the absorption spectra from nearby galaxies are shifted towards the red end of the spectrum. This is the Doppler shift due to the relative velocity between

Earth and the galaxy. Red shift indicates that the source is moving away from the observer.

The shift measured, and hence the relative velocity, was found to be approximately proportional to the distance of the galaxy from the Earth. Hubble's law is:

$$v = Hd$$

where v is the relative velocity between Earth and the galaxy, d is the distance between Earth and the galaxy, and H is the Hubble constant.

The value of H is not constant since gravity affects the motion of the expanding universe. Its value lies between 15 km s^{-1} per 1×10^6 cy and 30 km s^{-1} per 1×10^6 cy.

4.3 Age of the universe

The universe is expanding, so if we extrapolate backwards, the galaxies eventually come together to form one mass — hence the idea of the Big Bang when the universe was formed.

Assuming galaxies move out with a constant velocity v:

$$v = \frac{d}{T_0}$$

where T_0 is the time since the Big Bang. Hence:

$$T_0 = \frac{1}{H}$$

Background microwaves

The Big Bang would have generated black-body radiation at very high temperatures. As the universe cooled, the black-body spectrum would have changed to one corresponding to a temperature of about 3 K today. Such a spectrum with a λ_{max} in the microwave region was discovered in 1965.

Refer back to Wien's displacement law.

4.4 The cosmological principle

The universe is homogeneous and isotropic at any instant in time.

This is true for the universe as a whole, except that galaxies cluster in several regions.

The critical mass of the universe

Gravity slows down the expansion of the universe. Whether the expansion slows down to a stop and reverses depends on the density of the universe.

$$\text{the critical density } \rho_c = \frac{3H^2}{8\pi G} \approx 1 \times 10^{-26}\ \text{kg m}^3$$

This density corresponds to about five hydrogen atoms per cubic metre.

Current estimates suggest that the density of the universe is close to the critical value, suggesting a flat universe.

ρ_c can be calculated by a similar method to the escape velocity of a projectile from the Earth (see Topic 8, Section 3.6):

$$\frac{1}{2}mv^2 = \frac{GMm}{R}$$

$$\frac{1}{2}m(HR)^2 = \frac{GMm}{R}$$

$$\rho_c = \frac{M}{\frac{4}{3}\pi R^3}$$

$$\rho_c = \frac{3H^2}{8\pi G}$$

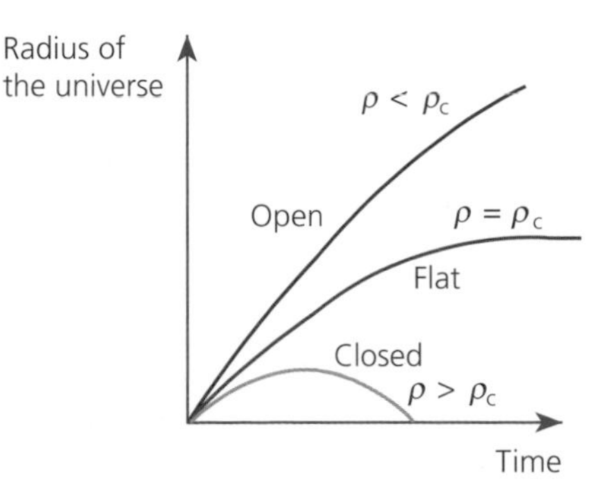

4.5 Quasars

Quasars are stellar objects that:

- are very bright and distant, and travelling at speeds up to $0.9c$
- have luminosities 100 times greater than local galaxies
- show strong emission lines rather than the absorption lines seen in nearby stars
- emit radio waves and X-rays
- exhibit short-term fluctuations in output, of the order of months, weeks or days

Quasars are thought to be high-energy sources concentrated in a small volume associated with active galaxies. Use the Hubble constant of $25\,\text{km}\,\text{s}^{-1}$ per 1×10^6 cy.

Worked examples

Q1 Galaxy NGC 379 has a velocity of $5500\,\text{km}\,\text{s}^{-1}$ away from Earth. How far is it away from Earth in parsecs? Use a Hubble constant of $25\,\text{km}\,\text{s}^{-1}$ per 1×10^6 cy.

$$d = \frac{v}{H} = \frac{5500 \times 10^3 \times 1 \times 10^6 \times 3 \times 10^8}{25 \times 10^3} = 6.6 \times 10^{16}$$

$$d = \frac{6.6 \times 10^{16}}{3.1 \times 10^{13}} = 2130\,\text{pc}$$

Q2 Calculate the age of the universe in years, given a Hubble constant of $25\,\text{km}\,\text{s}^{-1}$ per 1×10^6 cy .

Using $H = \dfrac{25 \times 10^3}{1 \times 10^6\,\text{cy}}$:

$$T_0 = \frac{1 \times 10^6 \times 3 \times 10^8}{25 \times 10^3} = 1.2 \times 10^{10}\ \text{years}$$

You should now know:

- **Oblers's paradox and Hubble's law**
- **how to calculate the age of the universe**
- **the Big Bang and background microwave radiation**
- **how to calculate the critical density of the universe**
- **open and closed systems**

TOPIC 11 Medical physics

1 Physics of the eye and the ear

1.1 Physics of vision

Note the refraction at the cornea and the lens.

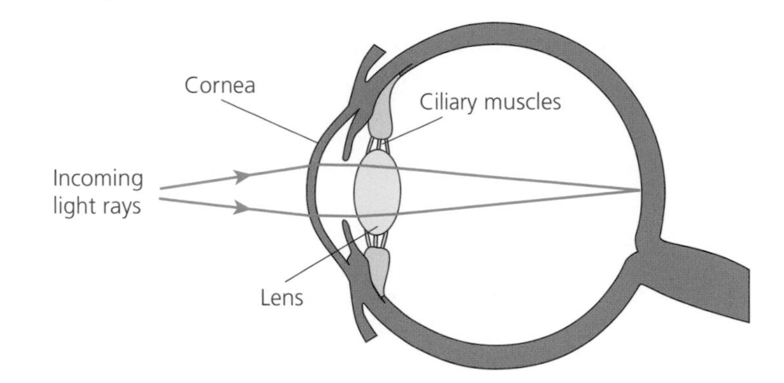

Spectral sensitivity

The eye can detect wavelengths from about 400 nm (violet) to 750 nm (red), with a peak in the green region at about 550 nm.

Defects

Myopia: short sightedness — the lens does not relax enough to focus distant objects.

Hypermetropia: long sightedness — the lens does not squash enough to focus near objects. This is the most common defect.

Astigmatism: the inability of the eye to focus on horizontal and vertical lines at the same instant. This is usually caused by an uneven surface on the cornea.

See Topic 7, Section 8.4, for the correction of these defects.

1.2 Physics of hearing

Frequency response: the ear can respond to frequencies between 20 Hz and 20 kHz. The upper limit decreases with age.

Threshold of hearing: the smallest sound intensity that can be heard — 1.0×10^{-12} watts m^{-2} at 1 kHz. The threshold varies with frequency, increasing at low and high frequencies.

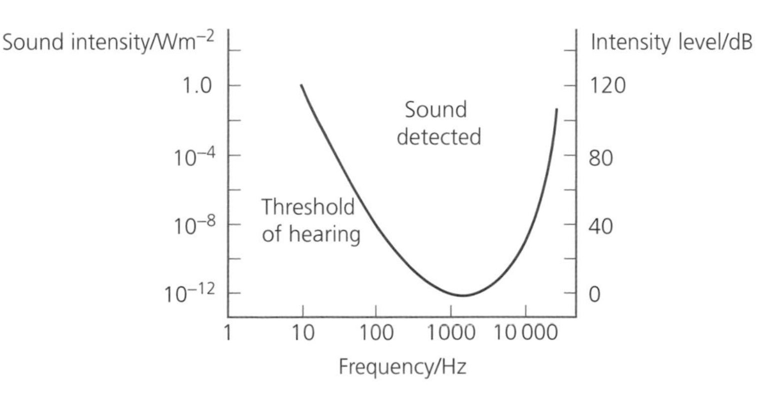

$$\text{sound intensity in decibels (dB)} = 10 \log \frac{I}{I_0}$$

where I_0 is the threshold intensity.

Sensitivity: the ability to detect a small fractional change ΔI in the intensity I.

The dBA scale is an intensity scale that responds in the same way as the ear with reduced sensitivity at high and low frequencies.

An intensity of $16 \times 10^{12}\,W\,m^{-2}$ is four times as loud as $2 \times 10^{12}\,W\,m^{-2}$.

Loudness: the subjective response of an individual to intensity levels. Loudness follows a logarithmic scale with respect to intensity levels.

$$\text{loudness} \propto \log \frac{I}{I_0}$$

Loss of hearing with age

As people age they suffer hearing loss in two respects:

- a reduction in the highest frequencies detected
- the threshold of hearing increases from $1.0 \times 10^{-12}\,W\,m^{-2}$

This is mainly due to a stiffening of the eardrum with age.

Worked examples

Q1 A person with poor sight can only see clearly objects placed between 2.2 m and 4.8 m from the eye. Calculate the power of the lenses required for objects placed at distances of 0.25 m and infinity from the eye.

An object at 0.25 m must produce a virtual image at 2.2 m:

$$\frac{1}{0.25} - \frac{1}{2.2} = \frac{1}{f} = 4 - 0.45 = 3.55D$$

An object at infinity needs to produce a virtual image at 4.8 m:

$$\frac{1}{\infty} - \frac{1}{4.8} = 0 - 0.21 = -0.21D$$

Q2 The threshold of hearing is a sound intensity of $1.0 \times 10^{-12}\,W\,m^{-2}$. If the threshold of pain is an intensity of $1.5\,W\,m^{-2}$, calculate the change in intensity in decibels.

$$\text{change} = 10 \log \frac{1.5}{1.0 \times 10^{-12}} = 122\,dB$$

Q3 An aircraft produces an intensity level of 105 dB at a distance 100 m away. Calculate the intensity level in $W\,m^{-2}$ at this point.

$$105\,dB = 10 \log \frac{\text{intensity}}{1 \times 10^{-12}}$$

$$\frac{105}{10} = \log \frac{\text{intensity}}{1 \times 10^{-12}}$$

$$3.16 \times 10^{10} = \frac{\text{intensity}}{1 \times 10^{-12}}$$

Hence:

$$\text{intensity} = 0.032\,Wm^{-2}$$

You should now know:

- **the physics of the eye and its spectral sensitivity**
- **the three main defects of the eye and their correction**
- **the physics of the ear**
- **the terms frequency response, threshold of hearing, sensitivity and loudness**

2 Biological measurements and imaging

2.1 ECG measurements

An electrocardiogram (ECG) is a graphic recording of the electrical activity of the heart. Electrodes are placed on the surface of the body to detect the electrical pulses used to activate the heart. The electrodes sense the electric field under the skin. There are two electrode configurations, unipolar and bipolar.

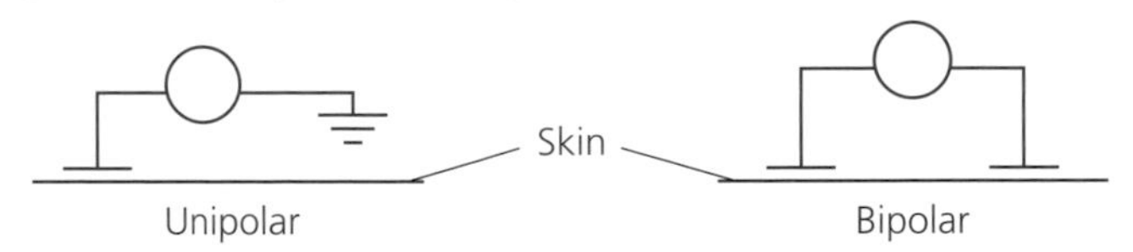

The heart

The heart can be thought of as a simple double-pump system controlled by electrical signals.

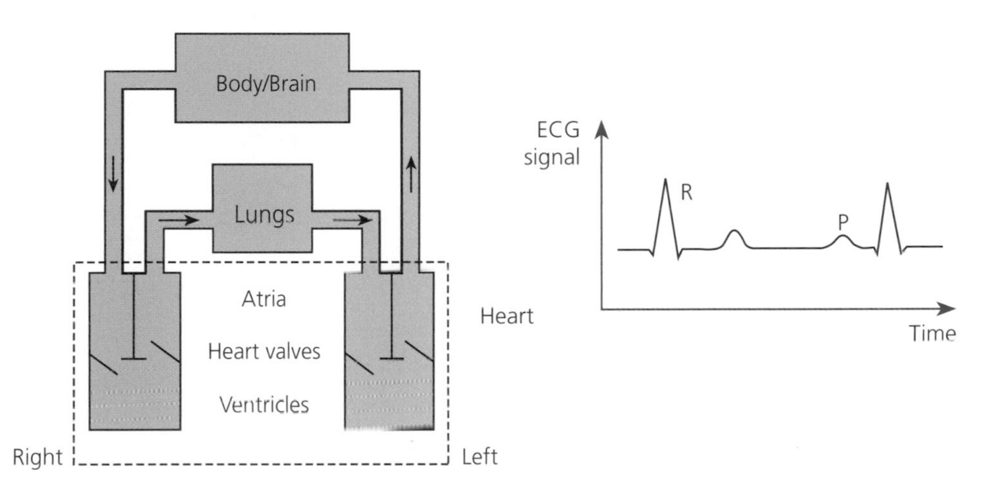

When the heart muscle relaxes, deoxygenated blood from the body flows into the right atrium; oxygenated blood from the lungs flows into the left atrium. The P wave occurs when the atria contract, forcing blood into the two ventricles. The R wave occurs when the atria relax and the ventricles contract, forcing blood from the right ventricle into the lungs and from the left ventricle into the body and brain.

Electrode positions

The heart is a three-dimensional object and electrodes placed at different positions pick up different signals. The standard positions are given by Einthoven's triangle:

- lead I, right arm to left arm
- lead II, right arm to left leg
- lead III, left arm to left leg

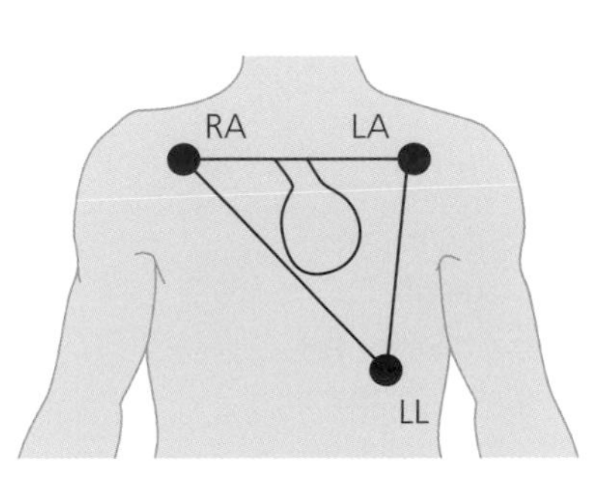

2.2 Ultrasound imaging

Ultrasound waves are very high frequency sound waves, greater than 20 kHz, generated by applying high-frequency voltages to a piezoelectric crystal, which expands and contracts as the voltage is applied.

Piezoelectric transducers

Such devices use frequencies in the MHz range up to 600 MHz.

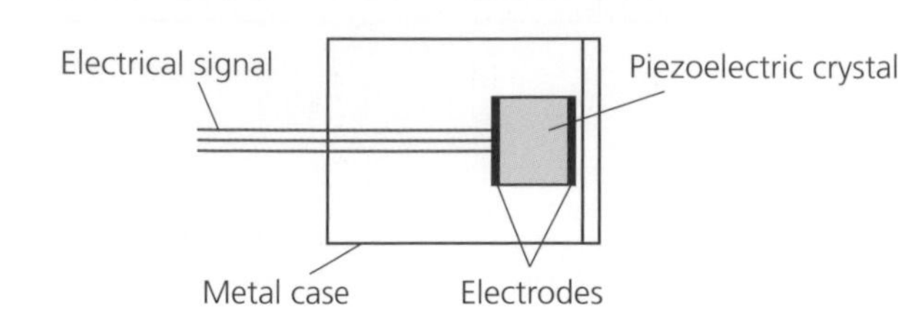

2.3 Acoustic impedance

Sound waves suffer reflection and refraction at the boundary between two different media.

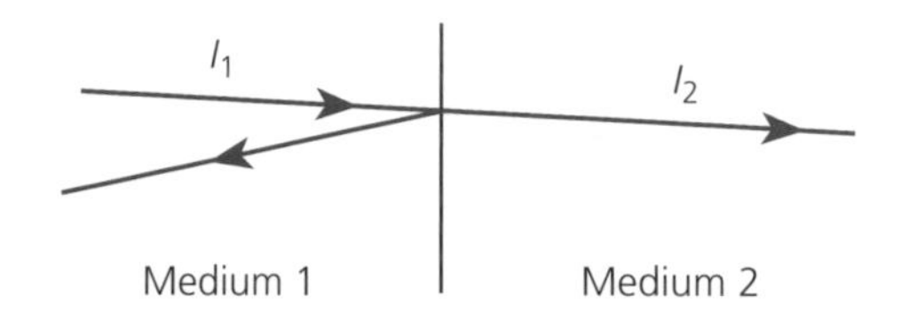

The ratio I_1 to I_2 depends on a property of the two media known as the specific acoustic impedance, Z. The acoustic impedance of a medium is proportional to the speed of sound in the medium and its density.

$Z_{air} = 430\ \text{kg m}^{-2}\ \text{s}^{-1}$
$Z_{bone} = 6 \times 10^6\ \text{kg m}^{-2}\ \text{s}^{-1}$
$Z_{tissue} = 1.6 \times 10^6\ \text{kg m}^{-2}\ \text{s}^{-1}$

$$\text{reflection coefficient } \alpha_r = \frac{(Z_2 - Z_1)^2}{(Z_2 + Z_1)^2} = \frac{I_1}{I_2}$$

2.4 Ultrasonic absorption

The intensity of the acoustic wave decreases exponentially with distance:

$$I = I_0 e^{-kx}$$

where k is the absorption coefficient and x is the distance travelled.

2.5 Ultrasound scanning

A scan: a short burst of ultrasound is transmitted into the body of the patient by a piezoelectric transducer. Sound waves are reflected from the boundaries and picked up by the same transducer now acting as a receiver. The signals are then displayed on an oscilloscope.

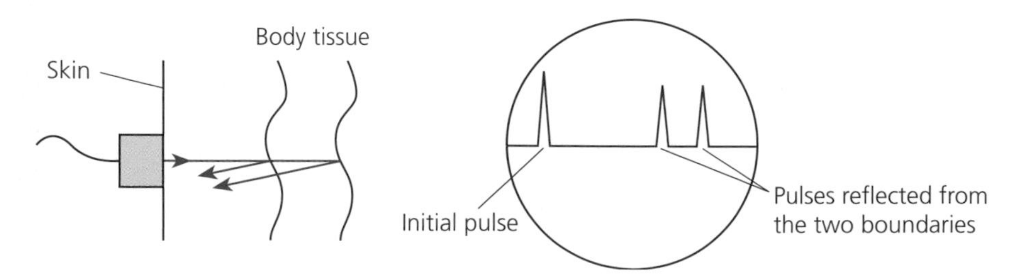

Ultrasound has the advantage of minimal health risk to patients compared with other imaging methods. It is used in fetal monitoring for this reason.

Coupling medium: the acoustic impedance between skin and air is very different and a large reflected signal is obtained at the boundary. A gel is used as a coupling medium to increase the transmitted signal to the patient.

B scan: a process in which the transducer head rotates through a small angle to produce a two-dimensional image.`

Worked examples

Q1 Calculate the fraction of sound reflected by an air–tissue interface and a tissue–bone interface.

$$\text{air–tissue} = \left(\frac{1.6 \times 10^6 - 430}{1.6 \times 10^6 + 430}\right)^2 = \left(\frac{1.5996 \times 10^6}{1.6004 \times 10^6}\right)^2 = 1$$

$$\text{bone–tissue} = \left(\frac{6.0 \times 10^6 - 1.6 \times 10^6}{6.0 \times 10^6 + 1.6 \times 10^6}\right)^2 = \left(\frac{4.4}{7.6}\right)^2 = 0.58^2 = 0.34$$

Q2 The absorption coefficient for air is $120\,\text{m}^{-1}$. Calculate how far an ultrasound wave will travel in air before its intensity is reduced by half.

$$\frac{I_0}{2} = I_0 e^{-120x}$$

$$0.693 = 120x$$

$$x = 5.7\text{mm}$$

You should now know:

- **how ultrasound is generated**
- **acoustic impedance and absorption**
- **the difference between A and B scans**
- **the need for a coupling medium**

3 *Laser: light amplification by stimulated emission of radiation*

Coherent means that all the waves are in phase.

The laser is a device that produces an intense beam of coherent radiation concentrated on a very small area of a few hundredths of a mm^2.

3.1 Helium–neon lasers

Helium atoms in the tube are exited by the large DC voltage. As they decay, energy is transferred to neon atoms.

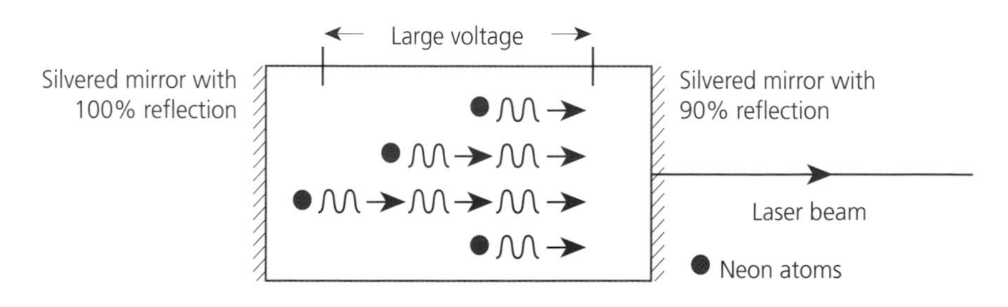

Excited neon atoms emit photons of energy 1.96 eV, light of wavelength 633 nm.

The excited neon atoms in the discharge tube can be stimulated to emit a photon as another photon passes. These two photons are in phase. The two photons then stimulate the emission of two further photons. The process is repeated down the tube. All the photons are in phase.

Care is required when using high-powered lasers because the intense energy concentrated on a small area can cause scarring of the tissue, particularly if the beam enters the eye.

3.2 Laser diodes

When a large current is passed across a pn junction, laser light is generated, which travels along the junction and is reflected at the two surfaces.

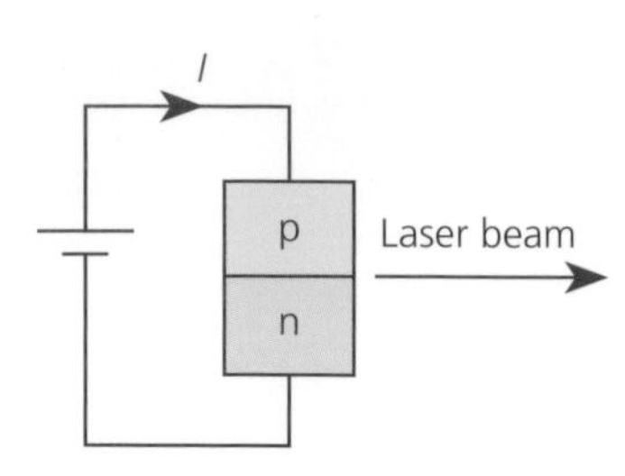

Uses of lasers

- Laser scalpels: lasers can be used as surgical knives with the advantage that the heat seals the skin, thus preventing bleeding.
- Pulsed lasers can generate shock waves to fragment bladder and gall stones.

3.3 Fibre optic cable

This is a fine glass cable with a surface of lower refractive index (μ) than the core. If light strikes the boundary at an angle greater than the critical angle, total internal reflection takes place and the light travels down the cable with very little energy loss.

The outer layer can also have a graded refractive index.

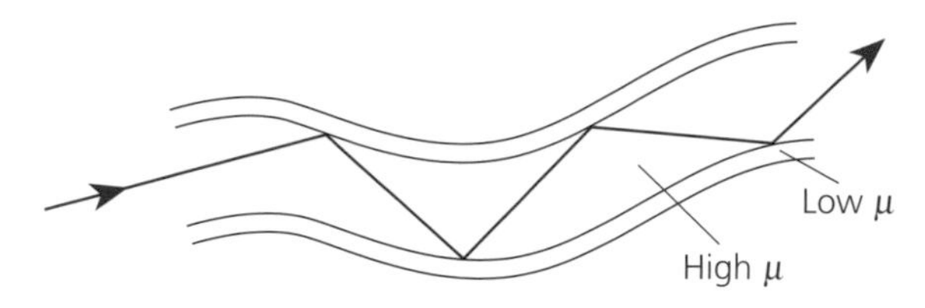

The **endoscope** is a device that uses bundles of optical fibres split into two groups, one for illumination, usually with a laser light source, and the second group to obtain the image.

You should now know:

- **the physics of the helium–neon and solid-state laser**
- **the design and operation of optical fibres**
- **their use in medical imaging and surgery**

4 *X-ray imaging methods*

4.1 Production of X-rays

The anode gets very hot and modern X-ray tubes use cooled rotating anodes.

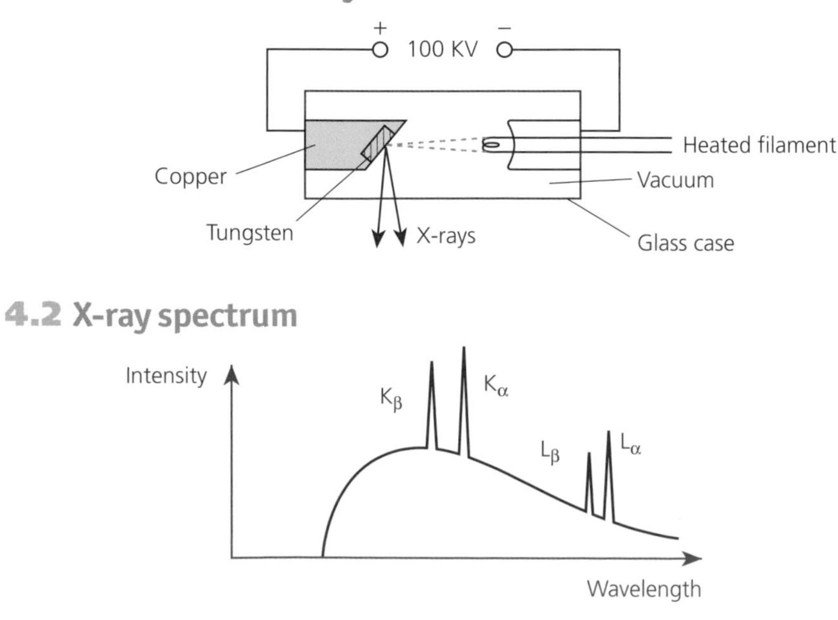

4.2 X-ray spectrum

The spectrum consists of two parts:

- a continuous spectrum produced when electrons slow down in the metal target, converting the loss of energy into X-ray photons
- a line spectrum produced when the electrons in the beam collide and eject electrons close to the nucleus of the target atoms; the subsequent rearrangement of electrons in the energy levels close to the nucleus gives rise to the emission of X-ray photons

4.3 Absorption of X-rays

The same equation applies to the absorption of γ-rays.

Lead is a good absorber of both X-rays and γ-rays.

Intensity decreases exponentially with distance:

$$I = I_0 e^{-\mu x}$$

where μ is the absorption constant and x is the distance travelled.

μ is proportional to the density of the material through which the X-rays pass.

Half-value thickness: the thickness of material required to reduce the X-ray intensity to half its initial value.

4.4 Image detection

With phosphor screens, 25% of the photons produce an image.

Shadow image: X-rays pass through the patient onto a photographic film (AgBr), which must be developed. A poor image is obtained with this method as only 2% of the X-ray photons produce an image in the film. The quality can be improved by placing the film between two fluorescent screens, which emit light when struck by X-rays.

Fluorescent screens: the patient is placed between the X-ray source and a fluorescent screen.

The technique uses the property of photo-stimulated luminescence.

Digital methods: when struck by X-ray photons, a phosphor plate captures a latent image. The plate is scanned by a laser beam, which releases the latent image as light photons. These are detected by a photo-multiplier. A digital image is obtained as the laser scans across the plate.

Contrast and sharpness

Contrast is lost because scattered X-rays reach the photographic plate causing fogging. These unwanted X-rays can be removed by placing grids in front of the detecting plate.

Sharpness is limited by the size of the heavy metal anode. It can be improved by using an anode with a small area or by using a metal plate to reduce the area through which the X-rays pass.

The complete arrangement

The aluminium filter removes low-energy X-ray photons, which cannot pass through the patient but would contribute to the radiation dose received.

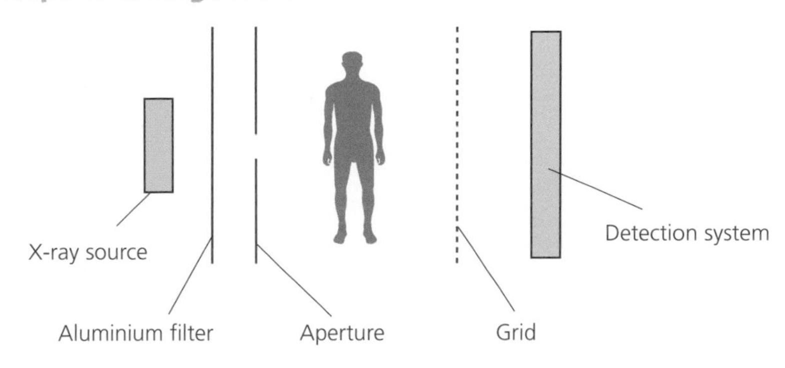

4.5 X-ray computed tomography (CT)

This is a scanning technique that enables a three-dimensional X-ray image to be generated.

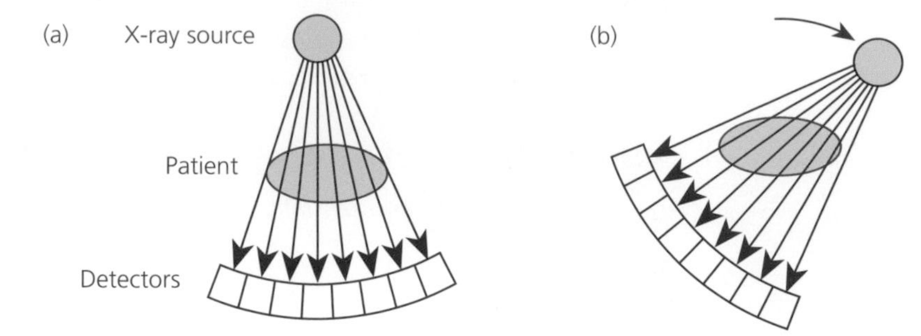

A broad X-ray beam passes through a thin cross-sectional slice of the patient. Detectors record the various intensity values and generate absorption values along adjacent rows (a).

The X-ray tube and detectors are rotated through an angle and the process is repeated (b).

The patient is fed through the detection system to obtain images of consecutive slices.

This is repeated for various angles and all the data for a slice of the patient are stored in a computer.

The following computer program then generates a two-dimensional X-ray image of the slice:

- the slice is divided into small volume elements called voxels, and an initial absorption value is assigned to each voxel
- the theoretical absorption values are adjusted repeatedly until the calculated absorption values along the various rows match the measured absorption values
- a final two-dimensional image is obtained
- this is repeated for consecutive slices to generate a three-dimensional computer image

Worked examples

Q1 The voltage across an X-ray tube is set at 2.5 kV. Calculate the maximum wavelength of X-rays obtained with this tube.

The maximum is obtained when 2.5 keV electrons are stopped.

maximum energy = $2.5 \times 10^3 \times 1.6 \times 10^{-19} = 4.0 \times 10^{-16}$ J

$$\text{energy} = 4.0 \times 10^{-16} = \frac{hc}{\lambda} = \frac{6.63 \times 10^{-34} \times 3 \times 10^8}{\lambda}$$

$\lambda = 5.0 \times 10^{-10}\,\text{m} = 0.5\,\text{nm}$

Q2 The intensity of a beam of γ-rays is to be reduced by a factor of 2000. Calculate the thickness of lead required if the absorption coefficient for lead is 57 m^{-1}.

$$\frac{I_0}{2000} = I_0 e^{-57x}$$

$7.6 = 57x$

$x = 0.13\,\text{m}$

You should now know:

- **the production of X-rays and the X-ray spectrum**
- **absorption and half-value thickness**
- **X-ray detection methods**
- **the design of a CT scanner**
- **how the final three-dimensional image is computed**

5 Magnetic resonance imaging (MRI)

The protons in the nuclei of hydrogen atoms possess spin, behaving as small magnets. When placed in a uniform magnetic field, they precess around the field in the same way as a gyroscope precesses in a gravitational field.

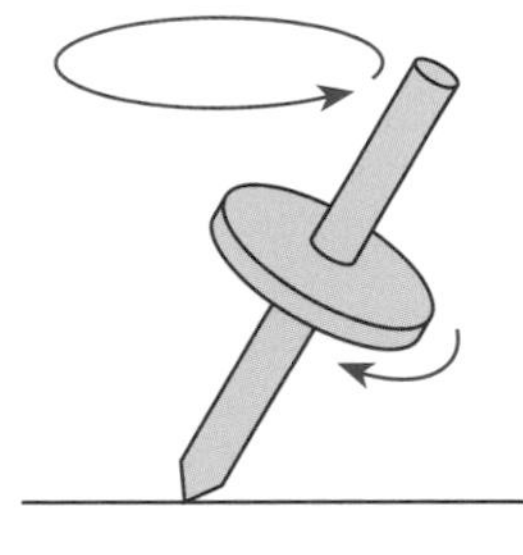

The precessional frequency is called the Larmour frequency. It is given by the equation:

$$f = \frac{\gamma B}{2\pi}$$

Note that small changes in B lead to small differences in f, the Larmour frequency.

where B is the applied magnetic field and γ is a constant. The value of γ is $2.7 \times 10^8\,\text{Hz}\,\text{T}^{-1}$, so that in a field of 1 T, f is 42.5 M Hz, which is in the radio frequency range.

When a short pulse of radiation is applied at this frequency, the protons gain energy and flip over to the reverse direction. This is a higher energy state and after a short time — the relaxation time — they return to the lower energy state, emitting energy at the same radio frequency as the applied frequency, which can then be detected.

5.1 MRI scanner

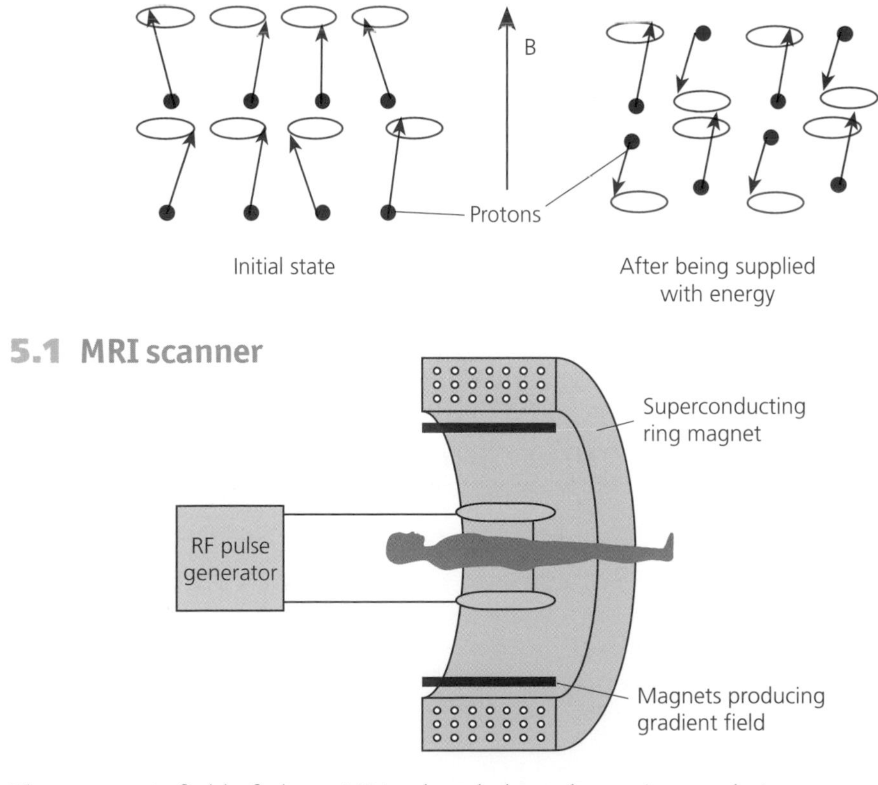

The magnetic field of about 1 T is placed along the patient and given a small gradient of about $20\,\text{mT}\,\text{m}^{-1}$.

A short pulse of radio frequency waves is applied to the patient, which excites hydrogen nuclei in a slice where the magnetic field and hence Larmour frequency matches the applied radio frequency. Hydrogen nuclei in this slice are excited to a higher energy state.

The same coil can be used for the application and detection of the radio frequency signals.

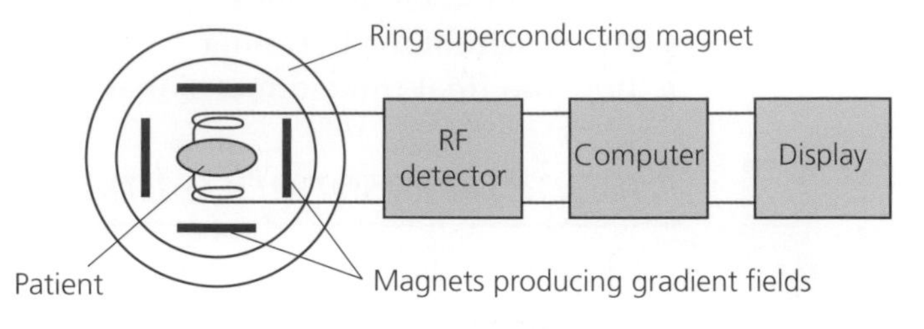

Two gradient magnetic fields at right angles are then applied across each slice. As the nuclei lose energy, flipping back to their lower energy state, radio frequency waves are emitted with slightly different frequencies depending on the position of the nuclei in the gradient fields. These signals decay with time depending on the local tissue environment.

By analysing different decay processes, information about different tissue environments can be obtained.

The detected radio frequency signal is then analysed by a computer into its various frequency components and a two-dimensional picture of a slice of the patient is obtained.

The frequency of the pulse of radio waves is changed slightly and protons in successive slices exited so that a three-dimensional image can be computed.

Worked example

Q1 A patient is placed in an MRI machine with a field of 1 T along the axis, which has a gradient of $20\,\text{mT}\,\text{m}^{-1}$. If slices are to be separated by 3 mm, what is the change in Larmour frequency required between slices?

change in B for a 3 mm slice $= 20 \times 10^{-3} \times 3 \times 10^{-3}$ T

change in frequency $= 2.7 \times 10^{8} \times 20 \times 10^{-3} \times 3 \times 10^{-3}$

$= 16.2$ kHz

You should now know:

- **the terms precession and Larmour frequency**
- **how energy is gained by a precessing proton**
- **the design of an MRI scanner**
- **how the final three-dimensional image is computed**

6 *Gamma cameras*

6.1 Gamma cameras

The scintillating crystals emit light when struck by a gamma ray. This light pulse is converted into an electrical signal by the photomultipliers.

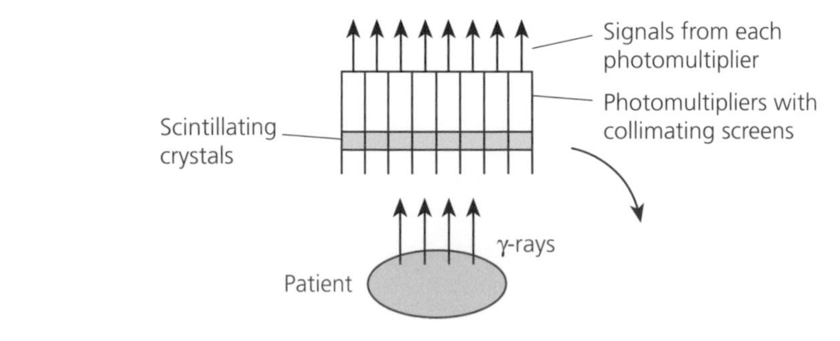

The resolution is lower than with CT scanning but gamma cameras have the advantage that a particular organ can be targeted and its function monitored over time.

The patient is given a suitable gamma emitter, which is absorbed by the organ under investigation. The patient is then placed under a gamma camera.

Emitted gamma rays generate electrical pulses in the various photomultiplier tubes and are counted and stored by a computer. The camera is then rotated around the patient and this is repeated for various angles of the camera. The stored results allow the computer to generate a three-dimensional image of the organ under investigation.

Organs larger than the dimensions of the camera require the patient to be moved through the camera and the above process repeated.

6.2 Positron emission tomography (PET)

The technique uses a positron-emitting radioactive tracer in which a proton in the nucleus decays into a positron. When the positron collides with an electron, they are converted into two gamma rays, which move off in opposite directions with an energy of 511 keV.

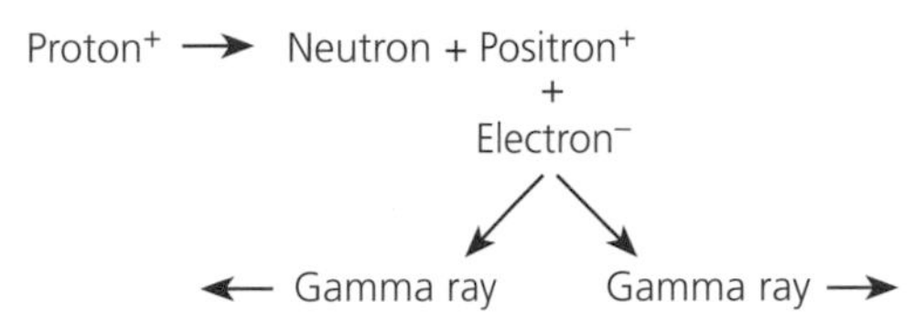

6.3 PET scanner

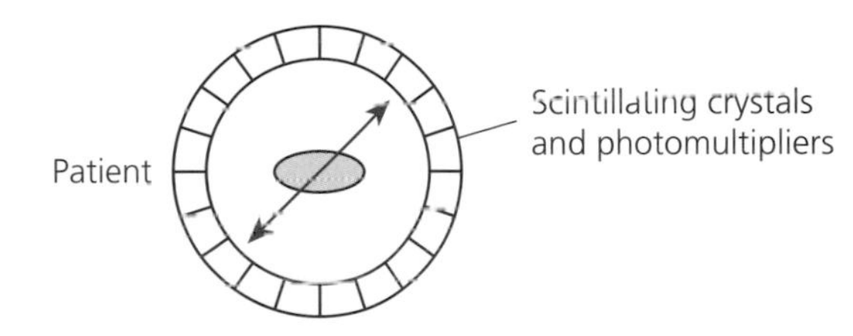

$^{18}_{9}F$ is a positron emitter with a half-life of 109 minutes.

The patient is given a suitable positron emitter, which is absorbed by the organ under investigation. The patients is then placed in the ring of detectors.

Gamma rays originating in the organ strike two detectors on opposite sides of the ring simultaneously, generating two pulses. They are thus distinguishable from random gamma ray pulses, which only generate single pulses.

A computer records simultaneous pulses from the detectors and a two-dimensional image of the positron emission is constructed.

As the patient is fed through the circle of detectors, a thee-dimensional image can be constructed.

You should now know:

- **the design and operation of a gamma camera**
- **the process of positron emission**
- **the design and operation of a PET scanner**

TOPIC 12 Devices and analogue electronics

1 Devices

In p-type silicon the major charge carriers are positive charges; in n-type silicon the major charge carriers are negative charges.

1.1 The junction diode

A junction diode is a device constructed by joining together two types of silicon, p-type and n-type, to form a pn junction. When an external voltage is applied across the junction, current flows in one direction only.

The arrow indicates the direction of conventional current when the diode is forward biased. V_B is the reverse breakdown voltage.

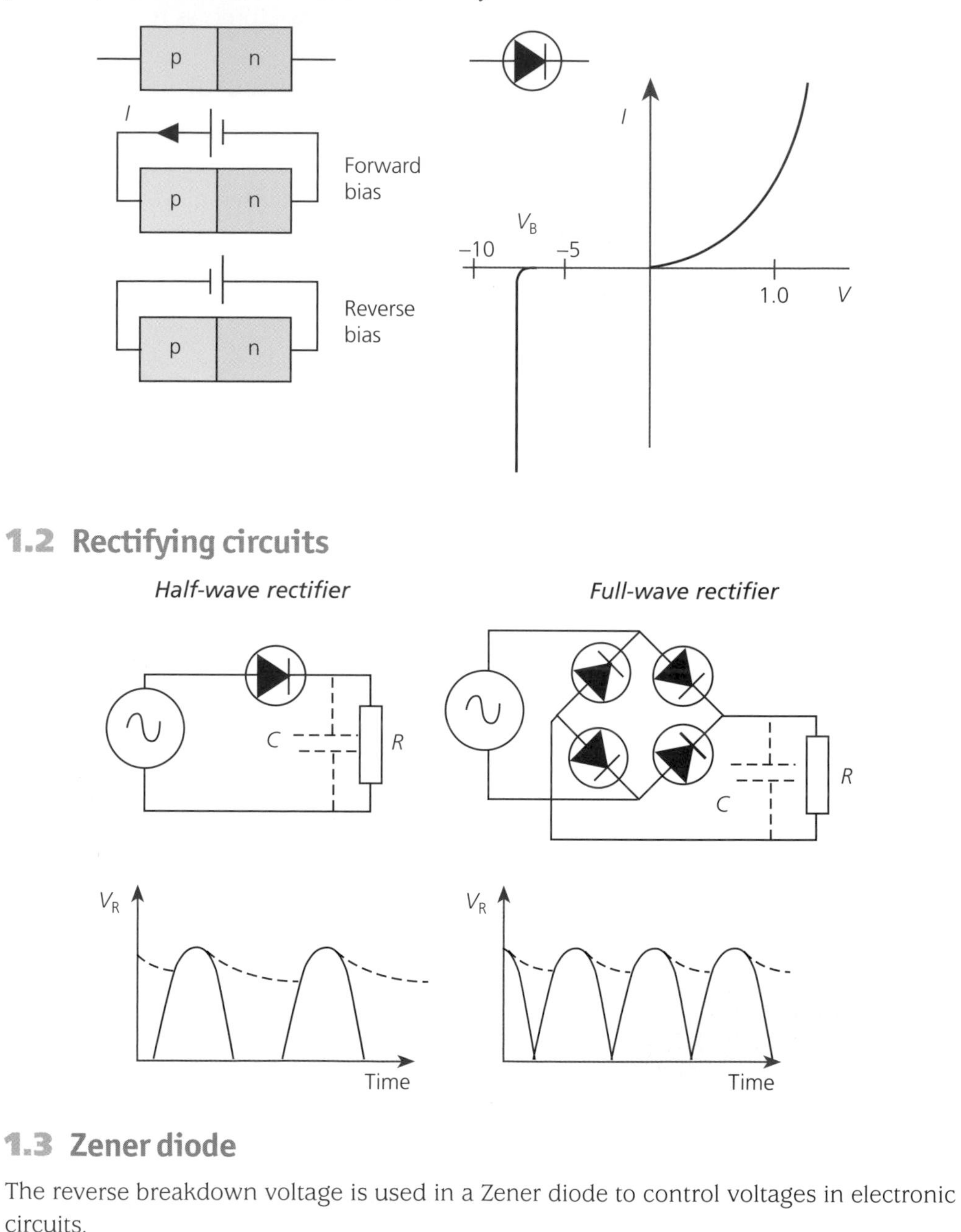

1.2 Rectifying circuits

Half-wave rectifier

Full-wave rectifier

In order to obtain true DC voltages, a capacitor is placed across the output. The capacitor charges up during part of the cycle and discharges through the load during parts of the cycle when the voltage falls.

1.3 Zener diode

The reverse breakdown voltage is used in a Zener diode to control voltages in electronic circuits.

In Zener diodes, V_B is in the range −6 V to −10 V. Avalanche diodes use a similar effect but have larger breakdown voltages — up to −50 V.

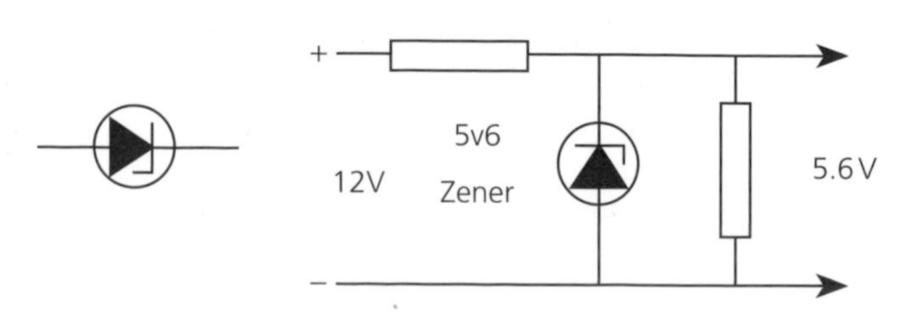

1.4 Light-emitting diodes (LEDs)

The LED converts electrical energy into light energy.

When current passes through a diode, light is emitted in the region of the junction as positive and negative charge carriers meet each other. They combine and release energy as light photons.

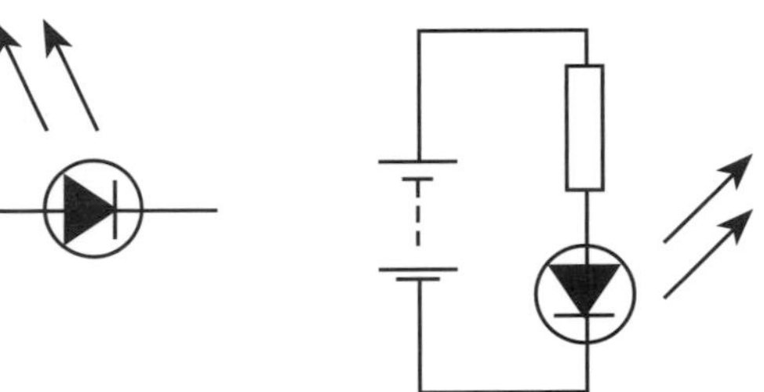

Light-emitting diodes use combinations of gallium, arsenic and phosphorus to obtain different colours in the visible spectrum:

- GaAs emits wavelength 900 nm — infrared
- GaAsP emits wavelength 650 nm — red
- GaP emits wavelength 560 nm — green

1.5 Solar cells

The solar cell converts light energy into electrical energy.

A solar cell is a pn junction with a very thin p-type layer. Light falling on the cell absorbed in the region of the junction creates positive and negative charge carriers. These charge carriers move to opposite sides of the diode, which act as the positive and negative plates of a battery. When connected to an external circuit, a current flows.

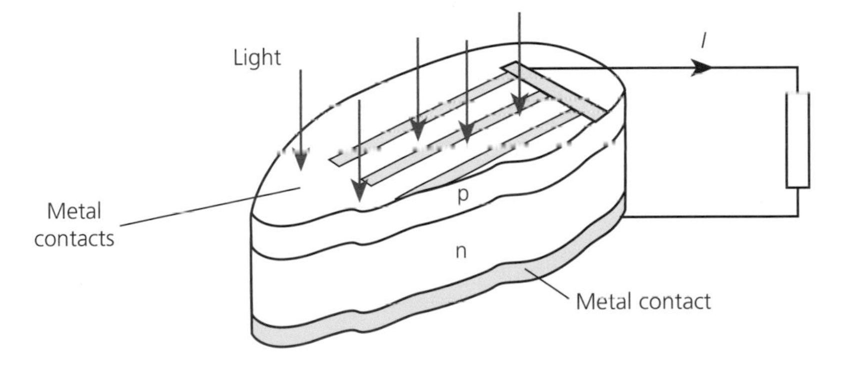

Worked examples

Q1 A current of 8 mA flows through a diode when the forward bias is 0.7 V. With a reverse bias of 0.7 V the current is 50 pA. Calculate the resistance of the diode in both cases.

$$R_{\text{forward}} = \frac{0.7}{8 \times 10^{-3}} = 87.5\,\Omega$$

$$R_{\text{reverse}} = \frac{0.7}{50 \times 10^{-12}} = 1.4 \times 10^{10}\,\Omega$$

Q2 A 10 V Zener diode has a burn-out current value of 20 mA. What is the maximum voltage that can be regulated with a series resistance of 2 kΩ?

voltage across 2 kΩ = $20 \times 10^{-3} \times 2 \times 10^{3} = 80$ V

maximum voltage that can be regulated = 80 + 10 = 90 V

Q3 The light photons in an LED are generated with an energy of 1.9 eV. What is the wavelength of the light emitted?

$$E = \frac{hc}{\lambda}$$

Hence:

$$\lambda = \frac{hc}{E}$$

$$\lambda = \frac{6.63 \times 10^{-34} \times 3 \times 10^{8}}{1.9 \times 1.6 \times 10^{-19}} = 6.54 \times 10^{-7}\,\text{m}$$

You should now know:

- **the characteristics of a junction diode**
- **reverse breakdown voltage and the Zener diode**
- **the layout of half-wave and full-wave rectifying circuits**
- **the operation of an LED and a solar cell**

1.6 The junction transistor

Transistors can also be constructed as pnp devices.

A junction transistor is constructed by placing two diodes back to back.

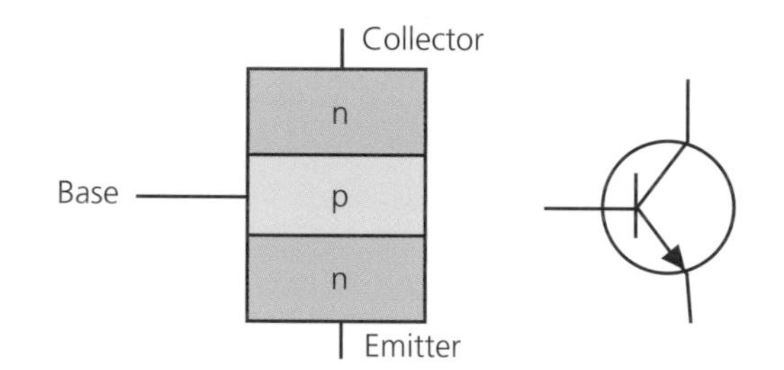

In an npn transistor, the base is made positive with respect to the emitter. The diode between the base and the emitter is forward biased. This allows a current, I_c, to flow from the collector to the emitter. A small current, I_b, in the base circuit controls a large current in the collector circuit.

The voltages are reversed in the circuits for a pnp transistor.

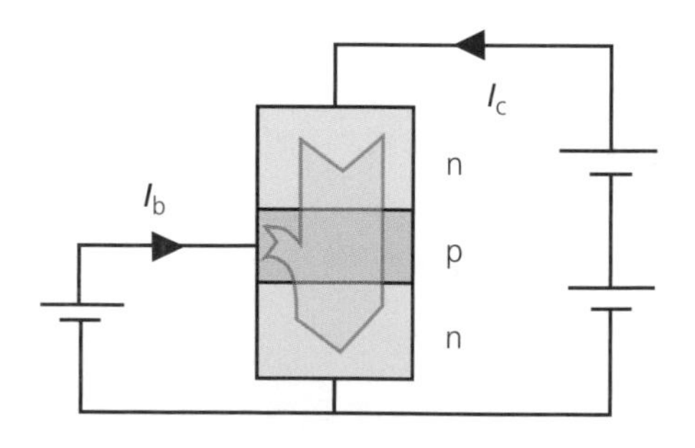

1.7 Switching circuits

Note that when the input signal rises the output falls.

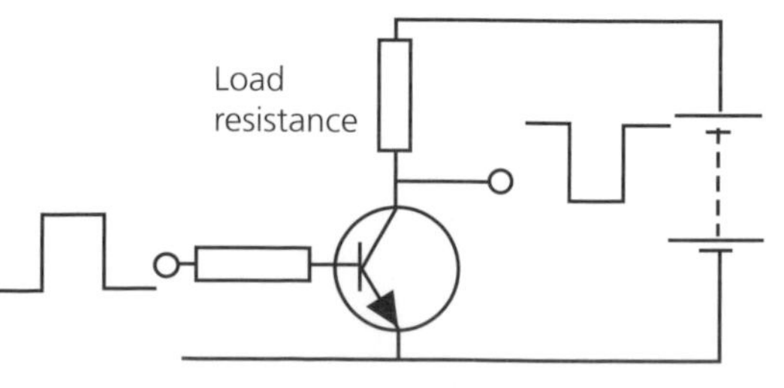

When a load resistance is placed in the collector circuit, the transistor acts as a switch.

You should now know:

- **the construction of a junction transistor**
- **its operation when suitably biased**
- **the circuit required so that it can be used as a switch**

2 Instrumentation: a sensing system

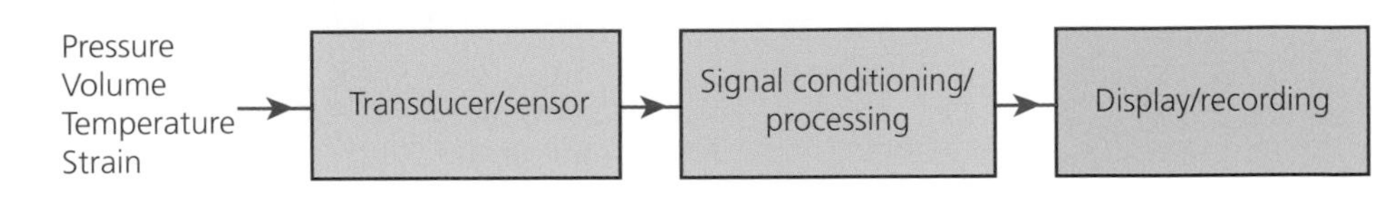

- The transducer converts the quantity being measured into an electrical signal.
- Signal conditioning modifies the signal in some way — amplification, filtering, modulation etc.
- Display/recording — the output is fed to a display/recording device (meter, LED display, chart recorder, computer etc.)

2.1 Transducers

Light-dependent resistor (LDR)

A thin layer of cadmium sulphide placed between two metal electrodes changes in resistance with the intensity of light shining on it.

Typical values of resistance are 1 MΩ in the dark to 100 Ω in sunlight.

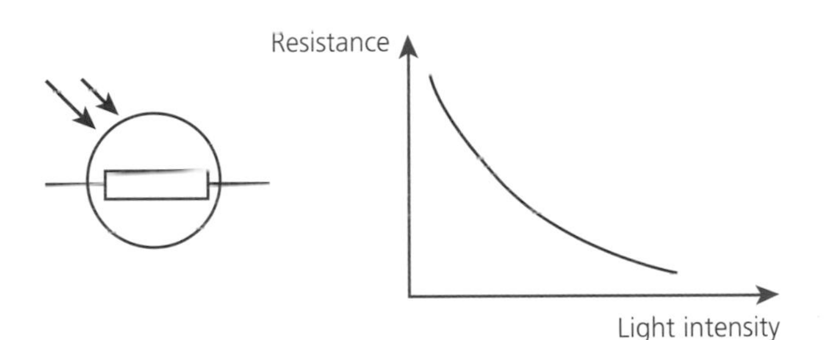

There is a non-linear relationship between light intensity and resistance.

Thermistor

A thermistor is a device constructed from oxides of various metals in which there is a significant change in resistance with temperature. The most common type is the negative temperature coefficient (NTC) thermistor, in which the resistance decreases as the temperature increases.

The resistance of a metal wire increases linearly with increasing temperature but the change in resistance is very small compared with a thermistor.

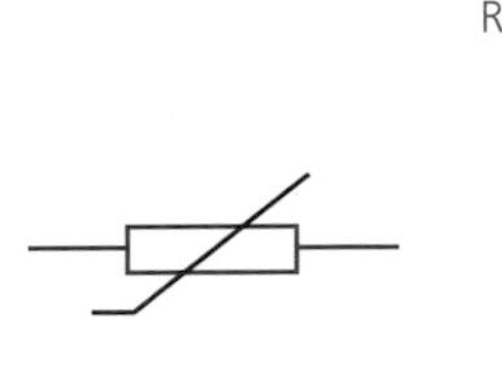

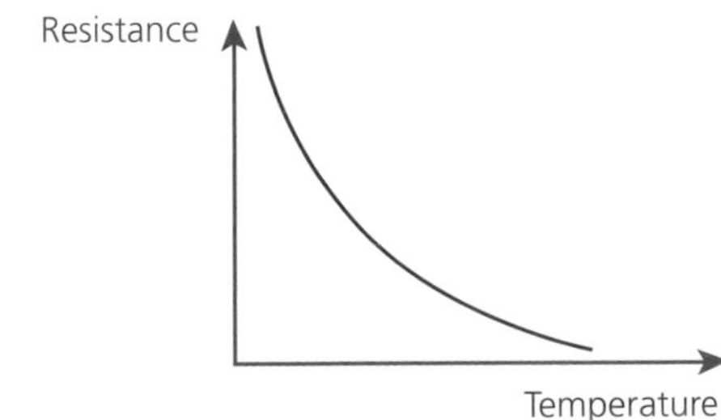

Note again that the change is non-linear.

Strain gauge

A strain gauge is a very thin wire fixed to a thin plastic rectangle, which changes in resistance when an extensive or compressive strain is applied.

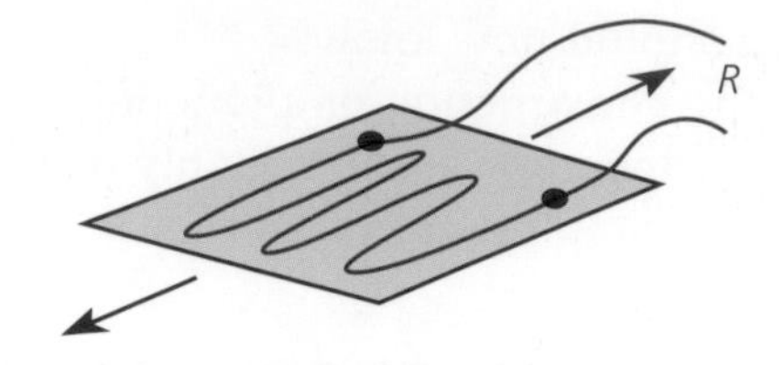

The resistance of the wire is given by:

$$R = \frac{\rho L}{A}$$

where L is the total length of the wire, A is the cross-sectional area and ρ the resistivity of the curve. The length increases under tension and the resistance increases linearly with ΔL.

In practice A decreases under tension to $A - \Delta A$ and R still increases.

$$\Delta R = \frac{\rho \Delta L}{A}$$

Piezoelectric transducer

This is a device that generates charges across the two faces of a crystal when compressed or expanded (see Topic 11, Section 3.2).

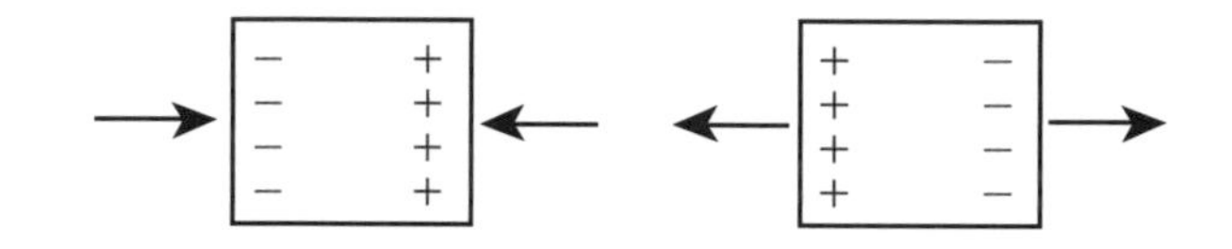

2.2 Generating an output voltage from resistance transducers

The change in resistance generated by an LDR or thermistor is converted to a voltage using a potential divider circuit.

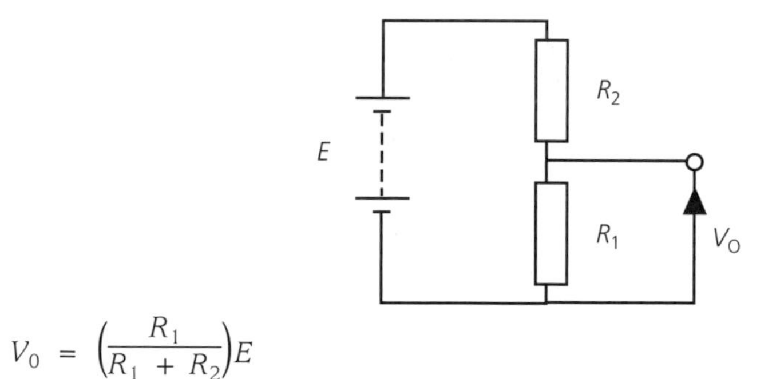

$$V_0 = \left(\frac{R_1}{R_1 + R_2}\right)E$$

Worked examples

Q1 An LDR has a resistance of 0.5 Ωm in the dark and 600 Ω in sunlight. If it is placed in series with a 1000 Ω resistor and a 12 V supply, what voltage is generated across the resistor in the dark and in sunlight?

$$V_0 = \frac{0.5 \times 10^6}{(0.5 \times 10^6 + 1000)} \times 12 = 12\,\text{V}$$

$$V_0 = \frac{600}{(600 + 1000)} \times 12 = 4.5\,\text{V}$$

Q2 A thermistor has a resistance of 3000 Ω at 40°C. What resistance must be placed in series with the thermistor and a 6 V supply to generate 5 V across the resistance when the temperature is above 40°C?

$$5 = \frac{R}{(R + 3000)} \times 6$$

$$5R + 15\,000 = 6R$$

$$R = 15\,000\,\Omega$$

Q3 A strain gauge is made from copper wire of resistivity $1.72 \times 10^{-8}\,\Omega\,\text{m}$. The total length of wire used in the gauge is 6.4 cm and the wire has a diameter of 0.056 mm. The gauge is subject to a tensile strain of 0.015. Neglecting any change in the cross-sectional area, calculate the change in resistance.

$$\text{strain} = \frac{\Delta L}{L}$$

$$\Delta L = \text{strain} \times L = 0.015 \times 6.4 \times 10^{-2}$$

$$\Delta R = \frac{\rho \Delta L}{A} = \frac{1.72 \times 10^{-8} \times 6.4 \times 10^{-2} \times 0.015}{\pi \times (0.028 \times 10^{-3})^2}$$

$$\Delta R = 6.7 \times 10^{-3}\,\Omega$$

You should now know:

- **the three parts of a sensing system**
- **the construction and operation of a range of transducers**
- **the operation of potential divider circuits**

3 The operational amplifier

Operational amplifiers are used to amplify the small voltages generated by most transducers.

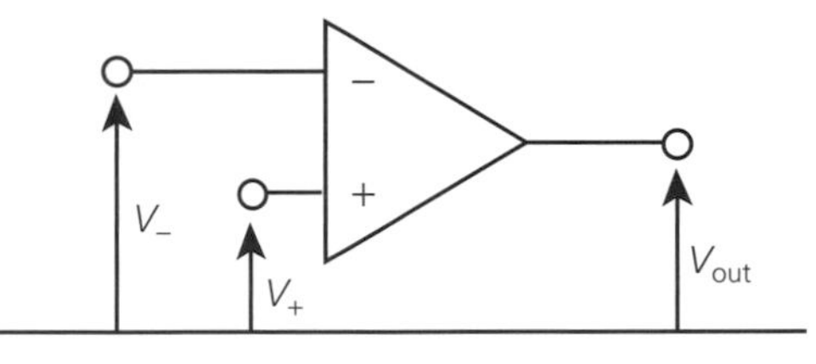

$$V_{out} = A_0(V_+ - V_-)$$

where A_0 is the open loop gain.

3.1 Ideal operational amplifier

An ideal operational amplifier has the following properties:

- infinite open loop gain
- infinite input impedance
- zero output impedance
- infinite bandwidth
- infinite slew rate

The gain of a real operational amplifier may be 10^6, not infinity; input impedance may be in the range 10^6–$10^{12}\,\Omega$, not infinity.

3.2 Operational amplifier circuits

Comparator circuit

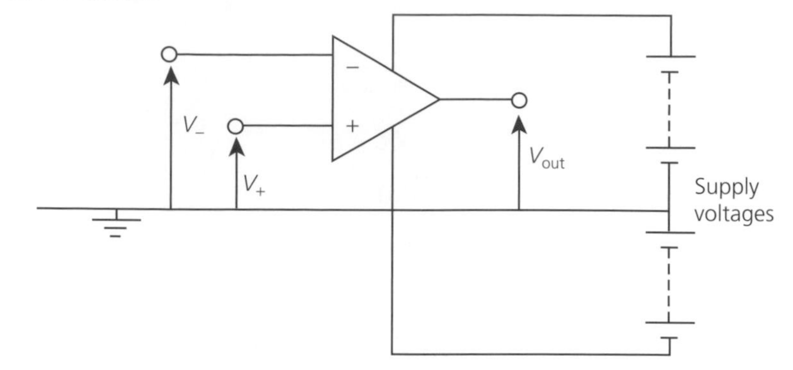

Connection to a bridge circuit

Any resistive transducer can be placed in one arm of the bridge to generate an output. Strain gauges are usually placed in bridge circuits.

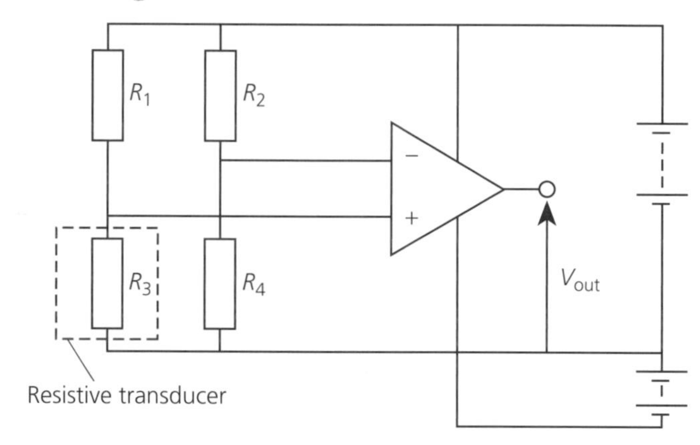

The problem with the above circuits is that, since A_0 is large, a small difference in voltage between V_+ and V_- will cause the output voltage to rise to the plus or minus supply voltage, and the amplifier saturates. The problem of saturation is solved using feedback.

3.3 Feedback and operational amplifiers

Negative feedback

A portion of signal from the output is fed to the inverting input.

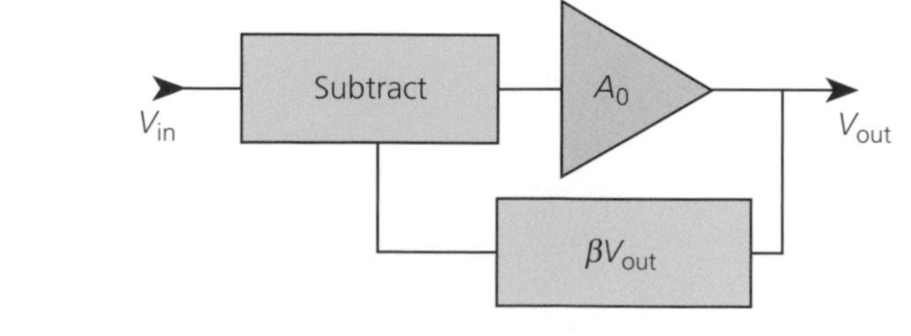

$$V_{out} = A_0(V_{in} - \beta V_{out})$$

$$\text{gain} = \frac{V_{out}}{V_{in}} = \frac{A_0}{(1 + A_0\beta)}$$

The gain is now less than A_0 but although the gain is reduced, the amplifier has other benefits:

- increased bandwidth
- less distortion
- greater stability

With positive feedback a signal is fed from the output to the non-inverting input, which can lead to instability. With careful circuit design, positive feedback can be used to build oscillating circuits.

Positive feedback

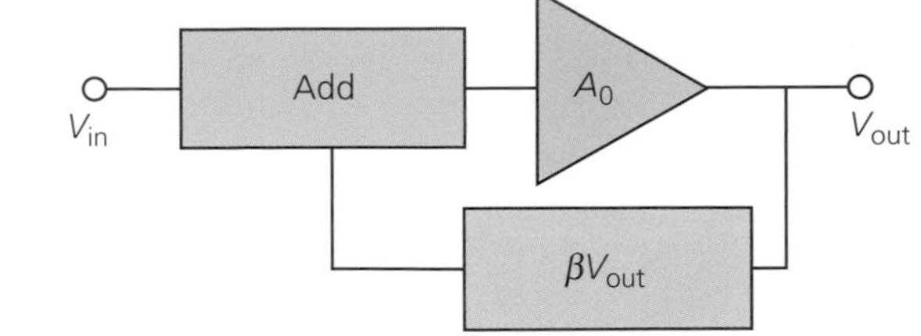

$$V_{out} = A_0(V_{in} + \beta V_{out})$$

$$\text{gain} = \frac{V_{out}}{V_{in}} = \frac{A_0}{(1 - A_0\beta)}$$

3.4 Inverting operational amplifier

R_f is the feedback resistance. R_{in} is the input resistance

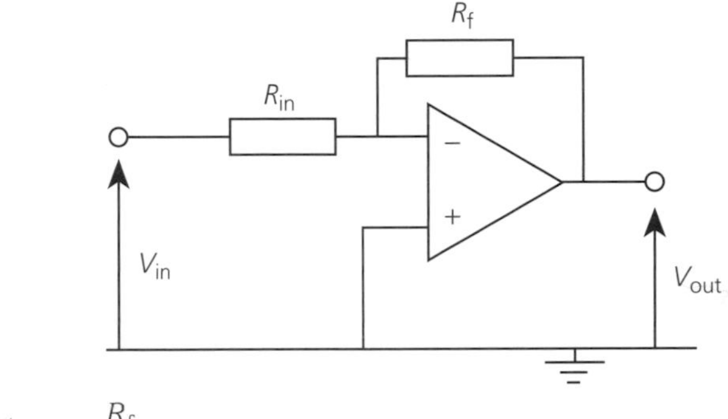

$$\text{gain} = \frac{V_{out}}{V_{in}} = -\frac{R_f}{R_{in}}$$

3.5 Non-inverting operational amplifier

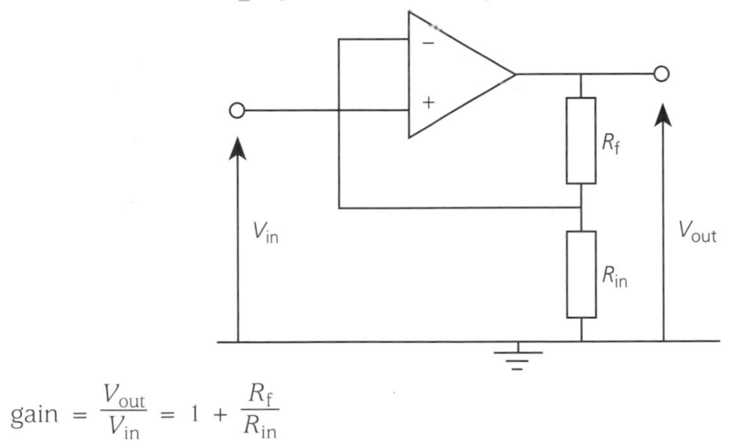

$$\text{gain} = \frac{V_{out}}{V_{in}} = 1 + \frac{R_f}{R_{in}}$$

3.6 Summing amplifiers

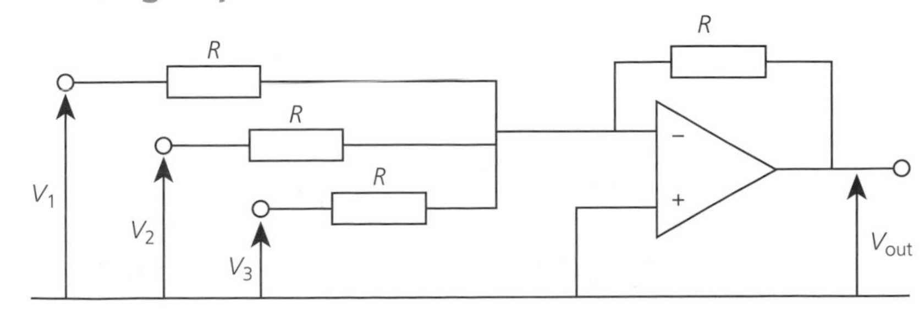

With all the resistances the same value:

$$V_{out} = -(V_1 + V_2 + V_3)$$

3.7 Relays

The output current that can be taken from an operational amplifier is very small (less than 25 mA). In order to switch larger currents, the output is often fed into a relay in which the output current feeds a magnetic coil. This pulls two contacts together and the larger currents can flow in the external circuit.

Relays are specified as one of two types in which the contacts are normally open (NO) or normally closed (NC) when no current flows through the magnetic coil.

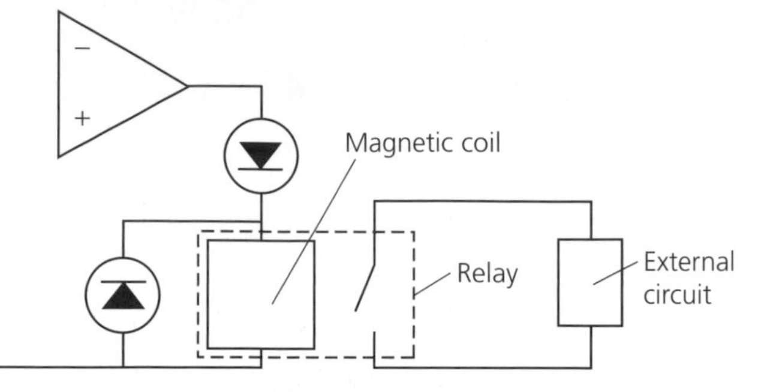

When the current in the magnet is switched off, a back EMF is generated and large currents could flow in the circuit, damaging the operational amplifier. Diodes are incorporated in the circuit as shown to ensure that the large current generated by the back EMF does not reach the output circuit of the operational amplifier.

Worked examples

Q1 An operational amplifier used in a comparator circuit has a gain of 1×10^5 and is connected to a ±15 V supply. What difference in voltage between the V_+ and V_- supply causes the amplifier to saturate?

$$V_0 = A_0(V_+ + V_-)$$

$$15 = 1 \times 10^5(V_+ - V_-)$$

$$(V_+ - V_-) = \frac{15}{1 \times 10^5} = 1.5 \times 10^{-4}\,\text{V}$$

Q2 An operational amplifier with negative feedback is supplied with an input voltage of 35 mV. If the input resistance is 1.5 kΩ, what feedback resistance would cause the amplifier to saturate when connected to a ±15 V supply?

$$\text{gain} = \frac{V_{out}}{V_{in}} = \frac{R_f}{R_{in}}$$

$$\text{gain} = \frac{15}{35 \times 10^{-3}} = \frac{R_f}{1.5 \times 10^3}$$

$$R_f = \frac{15 \times 1.5 \times 10^3}{35 \times 10^{-3}} = 6.4 \times 10^5$$

You should now know:

- **the properties of an ideal operational amplifier**
- **the layout of a comparator circuit**
- **the use of negative feedback with operational amplifiers**
- **the layout of inverting, non-inverting and summing amplifiers**
- **the need for relays in certain circuits**